The Face of the Past

The Face of the Past

Paul Jordan

B.T. Batsford Ltd London

First published 1983

Typeset by Servis Filmsetting Ltd, Manchester
and printed in Great Britain by
The Anchor Press Ltd
Tiptree, Essex
for the publishers
B.T. Batsford Ltd
4 Fitzhardinge Street
London W1H 0AH

ISBN 0 7134 4413 4

Contents

Acknowledgements

The author and publishers wish to thank the following for permission to reproduce the plates appearing in this book:

American School of Classical Studies, Athens 53
Argosy 23
BBC Hulton Picture Library 11, 33
British Museum 8, 9, 10, 12, 13, 14, 15, 51
British School of Archaeology, Jerusalem 30
Peter Clayton 37, 38, 39, 40, 47
Griffith Institute, Ashmolean Museum, Oxford 36
Hirmer Fotoarchiv, Munich 32
Manchester University Museum 43, 44, 45
Mansell Collection 42, 57
Phaidon Press Ltd 29, 35
J. Powell 46, 52
Reader's Digest 31
Schloss Gottorp Museum, Schleswig 54, 55
Staatlich Antikensammlungen und Glyptothek 48
Prof. William Watson, University of London 49
Weidenfeld (Publishers) Ltd 26
Werner Forman Archive Ltd 27, 41, 50, 56

The Illustrations

Time-scale of Events in Human Prehistory and History

30,000 years ago	The establishment of fully modern Homo sapiens sapiens and the first creation of works of art by the Old Stone Age hunters of the last Ice Age
9000 BC	The beginnings of the domestication of animals and the cultivation of plants in the ancient Middle East at the end of the last Ice Age
7000 BC	The spread of settled village life among the New Stone Age communities of the Middle East (Farming reached south-east Asia and south-east Europe by about 6000 BC, by which time it had also begun in Central America)
4000 BC	The first stages in the development of city-states in Mesopotamia
3000 BC	The beginnings of the Bronze Age in the ancient Middle East (bronze is an alloy of tin and copper, which had been worked cold in Turkey and Iran before 6500 BC); the development of systems of writing, first in Mesopotamia and shortly afterwards in newly-unified Egypt
2000 BC	The spread of civilisation around the ancient world (for example, with the appearance of bronze manufacture on Crete); the beginnings of civilisation in the Indus Valley and China
1500 BC	The coming to power of the Mycenaean Greeks in the Aegean; the establishment of the New Kingdom in Egypt; the first inklings of the Iron Age in Anatolia; the full establishment of the Bronze Age in Europe (where the final version of Stonehenge was completed at about this time)

1200 BC	The Greek sack of Troy and the Hebrew conquest of Canaan; the great extension of the Egyptian Empire under Ramesses II; the first civilisation of the Americas, with the Olmecs of Mexico
800 BC	The revival of the Greek city-states after a Dark Age of 300 years; the beginning of the Iron Age in Europe; the foundation (according to tradition) of the city of Rome in 753 BC
500 BC	The establishment of the Roman republic at about this time and the beginning of Greece's struggle with the Persians
300 BC	The division of the Empire of Alexander the Great among his generals and the consolidation of Roman power over all Italy; the flourishing of the Celtic Iron Age in Europe and the emergence of the Germanic tribes to the north
146 BC	Annexation of Greece by Rome
31 BC	Roman rule in Egypt and the end of the Roman republic with the supreme power of the first emperor, Augustus
AD 30	Possible date of the execution of Jesus in Jerusalem
AD 43	Roman conquest of Britain
AD 65	The execution of St Paul in Rome
AD 98	Approximate date of Tacitus' account of the habits of the Germanic tribes in his 'Germania'
AD 311	Ending of the persecution of the Christians in the Roman Empire
AD 391	Prohibition of Paganism in the Roman Empire

Human Evolution

as interpreted in this book

15,000,000 years ago	Ramapithecus, possibly the first of the hominid line ↓
4,000,000–1,000,000 years ago	Australopithecus, the first sure hominid (with a robust form off the direct line of human descent) ↓
1,000,000–200,000 years ago	Homo erectus, the first species of the genus Homo, including Java Man and Pekin Man ↓
250,000–30,000 years ago	Early Homo sapiens, including Neanderthal Man ↓
30,000 years ago till now	Homo sapiens sapiens, including Cromagnon Man and all the living races of today

Introduction

This book is about what human beings, including our remote 'ape-man' ancestors, looked like in the past. In particular, the book is about the facial appearance of our forbears as demonstrated by surviving physical evidence, actual bones and flesh if not flesh and blood. The emphasis here is upon the 'Face of the Past' rather than upon the stature of the past, for example, which in historical times at least had more to do with diet than with nature. At the same time, the emphasis is upon those comparatively rare and always striking instances when the very bones or mummified remains of long-dead human beings have chanced to survive down to our own day to show us what the people they represent looked like in life – rather than upon the abundant works of art which, in all ages and among virtually all human communities since the end of the last Ice Age, have illustrated more the way people saw themselves than the way they actually were. This book does, from time to time, glance at some of the pieces of representational art in which our ancestors have left us a picture of the way they liked to think of themselves: but only when, as in the very earliest manifestations of art in the Old Stone Age, the productions of the artists compare or contrast with the actual physical remains of their makers in an interesting way. Because the book is concerned with physical remains rather than with artistic depictions, details of costume and cosmetic are likewise frequently missing, though there are instances in which evidence for these things has vividly survived, as we shall see.

For the first several million years of human evolution, the only evidence of any sort for the progress (and even existence) of humanity consists of the products of human handiwork in imperishable forms (like flint implements) and the fossilised remains of our ancestors themselves. So rare are these fossil physical remains, over vast periods of time and through tremendous evolutionary transformations, that to chart the changing face of the human past by means of them is at the same time to chart the fundamental evolution of mankind. For this reason the first chapters of this book, that investigate the surviving faces of our ancestors up until the appearance of modern Homo sapiens, inevitably take on something of the epic colouring of the great story of human evolution itself. But this book's primary aim is to illuminate wherever possible the individual faces of humanity in the past that have chanced to come through to our times and arouse our curiosity: to that extent, the book is indeed a portrait-gallery of curious survivals of the past, set against the story of human evolution, prehistory and early history. In this context of curiosities, there is, moreover, a place even for the mistaken and downright fraudulent faces of the past, like the infamous Piltdown Man of allegedly prehistoric vintage and, perhaps, the figure on the 'Turin Shroud' in more recent historical times.

Since the arrival of fully modern sorts of human beings, during the last Ice Age, and especially since the beginnings of farming and settled living approximately 10,000 years ago, there have been no startling changes in the general appearance of the human face all over the world. There are marked racial, and for that matter individual, differences in the faces of current humanity, but (outside of the pathological) we do not nowadays meet with anything so strange as the faces of Java Man and Neanderthal Man and have not done so for some 30,000 years or more. And so the chapters of this book which deal with the chance survivals of human faces from the post-glacial past, since the end of the last Ice Age and the first beginnings of civilisations and written history, constitute more obviously a gallery of curiosities than do the chapters covering the long period of human evolution. Chance and the vicissitudes of human habits of burial have accounted for the physical survival of the later faces of the past: the Egyptian practice of mummification in particular has ensured the preponderant survival of ancient Egyptian faces over the rest of the peoples of the ancient world.

The apportioning of wall-space, as it were, in this gallery of the physical portraits of the past is, accordingly, weighted in favour of the ancient Egyptians (and to a lesser extent the Germanic tribes of Roman times, among others) on account of the chances of survival. These faces of the past undoubtedly possess curiosity-value for all the fortuity of their survival: but perhaps, in furnishing us with physical evidence for the appearance of certain human beings in past times when the taking of reliable portraits and photographs did not exist, they offer us something more of value to set alongside the numerous but never very reliable renderings of the human face in art. For, with the exception of some aspects of Egyptian art, the artistic depiction of humanity in the ancient world until the time of the Greeks was always 'distorted' by traditional imperatives away from the sort of naturalistic representation that the scientific temper of the twentieth century regards as 'real'. In any case, the haphazard survivals of human faces from the first farming communities of Jericho, from the predynastic and historical times of ancient Egypt, from the Iron Age bogs of Denmark and Germany, from the Imperial days of ancient Rome at Pompeii and Herculaneum are sufficient to spotlight graphic moments during the unfolding of the history of the ancient world. This book does not set out to present a potted history of mankind but, via the vivid faces of real men and women whose remains have come down to us, it hopes to bring forward occasional brightly-coloured individual portraits against a charcoal-sketched background of human history and prehistory.

Chapter One

The Evolution of Man

If the age of the Universe, according to the best estimates of our day, approaches something close to 15 billion years, then the utmost antiquity of man reaches back only to about one 3000th part of the total cosmic span. When our earliest known ancestors were gathering berries and raiding nests in East Africa some four or five million years ago, the Earth itself was already four and a half billion years old and life on Earth perhaps as much as three billions of years; the plant and animal kingdoms had diverged 600 million years before our ape-man forbears; the first fish had swum 450 million years before; the first animals to crawl on land had preceded our ancestors by 350 million years; the first mammals were 200 million years old; the first precursors of the Primates had appeared over 50 million years previously and the first apes and monkeys had come and gone by about 30 million years. When people question the Darwinian theory of evolution by random mutation and natural selection, they perhaps lose sight of the immense periods of time in which such mechanisms of change have been able to operate. This is not to say, of course, that other forces perhaps as yet unknown and perhaps unknowable have not also been at work.

Man and the great apes (of which the chimpanzee and the gorilla are his closest relatives) belong to a classification called the Hominoidea, a name coined from the Latin root for 'human being' and meaning 'human-like'. The Hominoidea appeared approximately 25 million years ago, and by about 13 to 15 million years ago it is considered possible that the ancestors of the apes (recorded by fossils found in Germany, Greece and Spain, as well as in Africa) and the ancestors of man (witnessed by fossils from Europe, Africa and India) had already diverged, with our own distant ancestors beginning to quit the tree-dwelling life for a regime in the grasslands beyond the forests. The possible ultimate human ancestor of more than ten million years ago is called Ramapithecus and it is important to remember that we are still dealing here with a very ape-like creature only some of whose features, like his teeth, are arguably beginning to point in the human direction. It is worth noting that studies of the blood chemistry and genetic material of the chimpanzees and gorillas have been deemed to place these primates closer to us than the zebra is to the horse. The notion is gaining ground, based on these haematological and chromosome studies, that the divergence between man and the great apes must have taken place at a considerably later time than was hitherto thought: in fact at about five or six million years ago, leaving Ramapithecus at about 13 million looking less inevitably like an ancestor of the human line.

After Ramapithecus, the fossil record becomes rather thin for upwards of ten million years: there have simply not survived to be found by us (so far) the fossils necessary to produce a detailed picture of the steps towards the next, more firmly identifiable human ancestor called Australopithecus.

Australopithecus (and more or less related forms) lived in southern and East Africa from some four million years ago (our oldest specimens) down to something more than a million years ago. His bones have been recovered in some quantity, with the occasional survival of bits and pieces like the limb bones and even digits, that are always rarer than skulls and teeth among fossils of early man. His bones render it possible to produce reconstructions of Australopithecus that make him out to be a plausible successor to Ramapithecus and ancestor of the fully human line: he was short of stature, with a small brain, but capable of bipedal locomotion (in other words, walking on two legs). There are at least two species within the genus Australopithecus, and anthropologists argue as to their precise status and classificatory nomenclature. It is, however, true to say that creatures of the general sort of Australopithecus constitute the ancestors of all the human species that have followed them. Among the Australopithecines appeared a momentous development in human evolution: the habit and tradition of tool-making. Other animals (including even birds) may be said to use 'tools' in the form of adventitious objects in the world about them to facilitate their acquisition of food: chimpanzees will sometimes use sticks to poke out what they want. But man and his direct ancestors are the only creatures to have systematically gone about the manufacture of tools for regular use. There is every reason to believe that tool-making marks a great intellectual advance in the evolution of man: it not only witnesses to an increase of intelligence among our ancestors, but it has almost certainly played its part too in fostering that increase in terms of intellectual and imaginative stimulation and the natural selection of better and better tool-making capacity. For tool-making conferred immense survival-benefits on its practitioners by enhancing their scope for the manipulation of the world about them, in catching and cutting up food in shapes and sizes and quantities beyond the means of non-tool-making creatures and working aspects of the environment into more useful forms, like shelters against the weather. The oldest tools, rather crudely modified pebbles, date back to about two and a half million years ago in Africa. Their broadly Australopithecine makers were able to build wind-breaks and eat (uncooked) meat.

'Australopithecus' is what zoologists call a 'genus': a classification that can include several different species. Modern man is zoologically known as 'Homo sapiens sapiens', which is to say the sapiens subspecies of the sapiens species of the genus Homo, or Man. The genus Homo evolved out of something like the genus Australopithecus over a period around two million years ago. The demarcation between the two genera is extremely difficult not only to identify in the fossil record, but also to define in ideal terms. But broadly speaking, it is relatively easy to assemble a picture of the traits and merits of the first clear-cut representatives of the genus Homo. By about one million years ago, a species of Homo called Homo erectus ('the upright-walking man', since those who named him in the last century – before the discovery of Australopithecus – believed him to be the first of our ancestors to walk on two legs) was living not only in Africa, but also in Asia and in Europe. He was living in a time of more pronouncedly variable climate than the world had seen before. It was in particular a time of such fluctuations in temperature (albeit over tens of thousands of years) that periods of warmer weather than today were beginning to alternate with periods of Ice Age when glaciers spread down over much of the northern part of Eurasia. There had been previous Ice Ages – for instance there was one about 280 million years ago – but never before had a creature anything like man faced their vicissitudes. The genus Homo went on to rise to the challenge very resourcefully as the well-nigh one million years of climatic variability developed.

Homo erectus sported a brain which, in his most developed form, reached about 1000 cc in size, twice the size of the brain of Australopithecus and three-quarters the size of the average modern man's. And in western Europe in particular, Homo erectus manufactured – over a very long period – a form of stone tool that signals a great advance on the chipped pebbles of the earlier African tool-makers. Recovered in great numbers from, among other places, the gravels marking previous courses of the English Thames and French Somme, well made 'hand-axes' (in actuality an all-purpose tool rarely, if ever, used for cutting down trees) point to the accomplished skills of the erectus flint-knappers in Europe and their long-term commitment to the repeated manufacture of a well-established and highly recognisable type of tool over literally hundreds of thousands of years. Homo erectus himself lasted for something like three-quarters of a million years and the hand-axe even outlived him, being last knapped in Europe probably as late as 50,000 years ago alongside other more sophisticated tool types. Of course, tools made of flint and other stones survive much better than those made in any other material available to ancient man and we know, from a very few survivals, that Homo erectus also made use of wood to produce spears, for example. In the South of France traces have been found of some sort of shelter made of wooden posts and stones, and from the same place and also from Spain and Hungary, has come evidence of the use of fire: a practice whose oldest indications occur at an erectus site near Beijing (Pekin) in China (though some recent evidence from Kenya may suggest the use of fire at an even earlier date, before one million years ago). We are dealing here with the *control* of fire acquired from forest blazes or the like, rather than the making of fire which comes much later. The improved tool and weapon technology of erectus and his use of fire involved him in big game hunting (including elephants in Africa) and the cooking of his kills. Life must have been enjoyed more abundantly by erectus than by his Australopithecus ancestors, with bigger and better-fed populations – though we are still talking of very small numbers by comparison with the modern world, with perhaps a few scores of thousands of erectus individuals in the world. The actual physical remains of erectus are few and fragmentary, but they chart in the Old World (for erectus was on his way to being the global species that man has since become) a progressive modernising of his features with increasing brain-size and gradually reducing face-size and ruggedness of skull. It was with Homo erectus, at least 500,000 years ago, that the rest of the skeleton below the skull achieved more or less its present form and stature: there has not been much change in the human skeleton apart from the skull for more than half-a-million years.

It is the development of the skull and the brain it contains that really charts the progress of human evolution: despite local variations, the characteristic curve of human evolution is the one that records growth of brain-size over the last million years. Not everything can be made of brain-size: there have been very talented men in modern times with erectus-sized brains and women, as everyone knows, have on average smaller brains than men to go with their, on average, smaller bodies. But it is clear that the progress of human culture and man's ever-increasing technological mastery of the world have been achieved to the accompaniment of an enlarging brain. The bigger brain, it seems certain, has been the unavoidable concomitant of better tool-making, enhanced exploitation of nature, the progress of invention and all the arts and sciences of man. Apes, despite the best efforts of some American researchers, do not speak and wield language. The use of language to communicate meanings and intentions and to foster social, shared activity must have played a crucial role in the evolution of man, and it is hard to believe that Homo erectus, with his highly traditional flint-knapping styles and his evidently communal hunting of his food and sheltering from the elements, did not employ some sort of speech and language, even though some researchers have tentatively concluded that all human ancestors prior to Homo sapiens sapiens possessed voice-boxes incapable of fluent speech.

After at least two major advances of the ice-sheets, during which man almost certainly retreated south to the latitudes of the Mediterranean, a long period of warmer weather ensued and lasted from perhaps about 400,000 to 250,000 years ago. During this time, Homo erectus evolved towards the earliest forms of Homo sapiens. At Swanscombe in southern England, in association with finely-made hand-axes and the bones of warm-loving animals, pieces of skull belonging to a more sapiens sort of man have been recovered. The skull, which is too fragmentary for complete reconstruction, is thick-boned but has been estimated to enclose 1300 cc of brain, very close to the modern average. 'Swanscombe Man' dates back to about 350,000 years ago. A very comparable skull comes from Germany, perhaps a bit younger and a bit smaller in capacity. At this stage our ancestors still possessed the heavy brow-ridges of erectus, but their cranial capacity, the shape of their skulls and the reduced size of their teeth were all heading in the sapiens direction.

Skulls from China, Greece and France continue the story – one from the South of France of about 250,000 years ago (when another cold phase of the Ice Age had set in) has been described as half-erectus and half-modern at a capacity of about 1200 cc, while a further French find, with a capacity of 1400 cc (above the modern average, though there have been cases in history of 2000 cc or more) may mark the beginnings of real Homo sapiens at about 150,000 years ago. It seems, by the way, that the large-scale occupation of caves as living-sites came in at about the same time.

Homo sapiens has so far come in two sorts or subspecies: anthropologists classify all living races of men as Homo sapiens sapiens; but earlier on there was another subspecies living particularly in Europe but found elsewhere in the Middle East and in Asia, called Homo sapiens neanderthalensis and commonly known as 'Neanderthal Man' (after the valley in Germany where his remains were found in the last century). There is quite a range in the physical features of Neanderthal Man, with the earlier specimens less 'Neanderthal-like' than the later ones, and the image of Neanderthal Man has suffered a bad press on account of an early reconstruction of him as a stooped and shuffling 'ape-man' based on an unrecognisedly but severely arthritic skeleton. Neanderthal Man was stocky and thick-set, and tended to a heavy face with marked brow-ridges, but his brain was large at 1450 cc (larger, in fact, on average than modern man's, though differently distributed with an apparently less well-developed frontal area, with whatever that may portend for the organisation of his mind). He was probably at least as presentable as the average hippie or biker and so might have passed unremarked in a modern crowd.

Neanderthal Man's stone tool-kit represents some sophisticated advances on what had gone before, with more varieties of tool-types and more specialisation. His scraping-tools point to the preparation of animal skins for the making of clothes – and indeed his survival in the north after about 70,000 years before the present, when the last great phase of climatic deterioration began, would have required weather-proof clothing and full provision against the cold winters. Neanderthal Man was the first sort of man to live in northern Eurasia in very cold conditions.

Among the Neanderthals we first observe the arrival of many traits that we associate with real humanity as against our animal origins. While there is no evidence of art among them, the Neanderthals did collect bits of ochre colouring material, perhaps for use in vanished body painting. From a Neanderthal site in Hungary comes a pebble with a cross incised on it. In a cave in Switzerland, Neanderthals had collected cave-bear skulls in a stone box: an apparent indication of some sort of cult. Neanderthals sometimes buried their dead ceremonially: in Russia a Neanderthal boy was buried with some goat horns; other Neanderthal burials show the provision of food and flint tools as though for a continuation of life beyond death; in one touching instance, a Neanderthal was buried under a carpet of flowers. Neanderthals appear to have occasionally practised cannibalism – which is almost always a sign of ritual rather than dietary concern.

Anthropologists do not agree as to the relationship between the Neanderthals and Homo sapiens sapiens: it is possible that the latter evolved out of something like the more generalised earlier Neanderthals and it is also possible that a biological effect called neoteny may have been at work to produce quite rapidly a more 'youthful' version of

Neanderthal Man – ourselves – to succeed him. Neoteny leads to sexual maturity at a more juvenile stage of development, and it is a fact that Neanderthal children's skulls are more like Homo sapiens sapiens than are the adults'. For that matter, so are the skulls of infant gorillas and chimpanzees. The appearance of fully modern man, in the shape of the well-known Cromagnons of Europe, has always been associated with a change in the flint tool-kit towards even more varied and sophisticated forms, but these too could have been developed out of the tools of Neanderthal Man.

At all events, by about 30,000 years ago, the technology of early man (particularly in Europe) had taken on new complexities, the type of man who made these tools had become fully modern in form, and the progress of those human traits to do with cult and belief, so tentatively foreshadowed among the Neanderthals, had greatly accelerated. The Cromagnons, as it is convenient to call them after an early find site in France, developed fine blades of flint as knives and scrapers, sharp-pointed carving and engraving tools and tools in bone and antler-like needles to sew with. Graves in Russia reveal them to have worn hide trousers and shirts against the winter cold and to have lived where they had no caves in skin huts. Their whole way of life across Europe and Asia during the final millennia of the Ice Age must have quite closely resembled that of some of North American Indian tribes of recent times and the Eskimos. In fact, the first crossings of the Bering Straits from Asia to the Americas, facilitated by the world-wide lowering of sea-level as a result of sea-water locked up in the great ice-caps of the Ice Age, was undertaken perhaps as early as 20,000 years before the present by groups of Homo sapiens sapiens hunters. During the following millennia both North and South America were populated by human groups obviously related to the Mongolian types of the Far East. The Eskimos are a much later phenomenon, however, of perhaps only a few thousand years BC, while the distinctive Indian cultures of the Plains were young when the Spaniards supplied them with the horse.

The most striking feature of the life of the Old World hunters of the last Ice Age is their art, a first spectacular expression of one of the most distinctively human accomplishments that stands on equal terms with anything achieved later by ancient Egyptians, Greek, Chinese, Indian or Renaissance artists. We have seen that Neanderthal Man took the first uncertain steps towards art in the use of ochre and very occasional patterned scratchings. Even chimpanzees show some glimmerings of aesthetic sensibility when a paintbrush is put in their hands. But the art of the Homo sapiens sapiens hunters is far ahead of all these fumblings and, lasting from about 35,000 to 10,000 BC, represents not only the first but the most enduring of man's artistic traditions.

A chronology of the cave art of the Ice Age – it is a very largely European affair and in fact overwhelmingly confined to the caves of south-central France, the Pyrénées and the north coastal region of Spain – is difficult to arrive at. But the best current estimates reckon that this art begins at about 35,000 years ago with small and portable decorated and coloured objects often of bone and ivory and some carved and painted blocks. The representational wall art seems to appear after about 20,000 years BC with some animal depictions and some symbolic forms, notably sexual and usually based on the human female genitalia. It has been suggested that sexual imagery underlies all the cave art, including the representations of animals, with a Freudian dichotomy between bulky 'female' creatures like bison and more elongated 'male' forms like horses and deer: at the very least, we can be sure that sexual imagery was a very strong component of this earliest art. A distinctive range of products are the so-called 'Venuses', small female figurines frequently of a grossly exaggerated character with huge breasts and thighs. When it came to showing human-beings, the Ice Age artists always tended towards caricature, as some of the male faces from the wall art also illustrate.

It seems that the urge to create art is a deeply rooted human drive; we should not look for any exclusive explanation of the meaning and purpose of the cave art. Nevertheless, there are some unmistakable pointers like the sexual imagery mentioned above, and it seems obvious that the principal motif of the developed cave art involves the animals which were surely the prey and the food supply of these hunters, as is attested by the bones found on kill sites and among the occupational débris of the caves. And so we think of hunting magic to increase the herds on which man fed at the same time as increasing his own fecundity. Among the not very numerous representations in all the cave art of what must be human beings, there is one from the French cave of Les Trois Frères that irresistibly suggests a medicine-man in his regalia, and we can reasonably infer that the classless society of the hunters did embrace chiefs and sorcerers. It is worth remembering that hunting societies do not work half as hard as the working sections of peasant and industrial communities do: it has been estimated that the average hunter does a 15-hour week. The work may well be intermittently strenuous and dangerous, and life expectancy may be curtailed, but hunters do have time on their hands – to think, to imagine, to create art and cult. No doubt the Ice Age hunter-artists involved themselves in a very complex web of beliefs and ritual whose only enigmatic survivals are the painted and engraved pictures on the walls of their caves and the little carvings and clay figures from the strata of their sites.

The art of the Ice Age hunters finally waned, after a span lasting much longer than the whole Western tradition since the earliest Egyptian art, in about 9000 BC when the environment of the hunters changed drastically with the retreat of the ice-sheets, the warming up of the climate and the spread of forests. Where once reindeer and mammoth had figured in the artists' repertoire, now the forest stag came to predominate; in the end the art dwindled via elaborately decorated bone tools and weapons into painted pebbles.

When the Ice Age ended about 10,000 years ago, the rich hunting life of the Cromagnons of Europe ended with it. Human occupation in the newly temperate latitudes of Eurasia did not cease, though no doubt some of the hunters followed their quarry north: a new way of life was evolved to suit the new conditions, with much fishing and fowling in marshlands and forest hunting of game. At the end of the Ice Age, man was a global species with representatives on all continents including the Americas and Australia. All over the world, the little communities of men and women were living a similar life of subsistence based upon hunting and gathering. But the stage was set for a momentous new development in human history: the development of the farming way of life, which was to produce surpluses and wealth, leisure and the civilised life for some, and drudgery for the majority in place of the often hard and uncertain, but classless and not too time-consuming, economy of the hunters.

Chapter Two

Australopithecus and Family

4,000,000–1,000,000 Years Ago

The great apes and man belong to a superfamily of primates called the Hominoidea: the orang-utan, the gorilla and the chimpanzee share the family of the Pongidae, while all living and extinct forms of human being belong to the family Hominidae. It is convenient to refer to all these sorts of men as 'hominids'. Recent work on the blood chemistry and genetic composition of both men and apes suggests that the hominids and the pongids shared a common ancestor perhaps as recently as five or six million years ago. That common ancestor was in truth the notorious 'Missing Link' who has achieved a sort of folk-hero status since Darwin's disciples popularised the notion of the descent of man from the animal world, in the last century. The early students of human evolution, working on a hypothetical time-scale that we now know was much too short, expected to find a creature of only a million years or so ago (perhaps even only hundreds of thousands) looking like a man-sized ape, with perhaps the beginnings of a nobler brow to mark the distinctively human enlargement of his brain and mental powers. The nineteenth-century German anthropologist, Haeckel, inspired the most vivid reconstruction of the sort of 'Missing Link' conjectured in his day: it was only informed conjecture since no bones had been found at that time on which to base a reconstruction.

The picture of human evolution emerging in our own time suggests that the identification of the elusive 'Missing Link' will be immensely more complicated than was previously thought. All the hominid fossil evidence available shows us that our early ancestors did not sport the big brains and noble brows that once seemed so essentially human. We must look for distinctive humanity at the other end of the human frame, with our remote ancestors' feet and habits of locomotion. Skeletons sufficiently complete to include foot and leg bones are rare, and so far missing altogether when we venture back in time beyond about four million years and try to identify the very earliest hominid divergence from the common hominid ancestor of apes and men.

Remains of the creature called Ramapithecus have been found in India and Pakistan, Kenya, Greece and Turkey. At first only jaws and teeth were discovered, which some workers thought were more human than ape-like, with the result that Ramapithecus was often considered as the earliest known hominid. But when some skeletal remains were found in 1977, they seemed more arboreal in character, suited to tree-living, than anything else. In 1979 a skull with jaws, face and part of the forehead was found in Pakistan: its molar teeth resembled human molars like those of previous specimens, but its canines and incisors turned out to resemble apes' teeth, with the whole strongly resembling the face of an orang-utan. Ramapithecus might thus represent the oldest known ancestor of the human line or the ancestor of both the hominid and pongid lines, or he might even be the ancestor of only the apes. In the absence of his leg bones, we cannot know his mode of locomotion and determine whether Ramapithecus had taken any steps towards the erect bipedal stride that only the hominids evince: nor can we know the size of his body for sure, but he was probably quite small. Nevertheless, in the fragmented visage of Ramapithecus, we can be reasonably sure that we are looking at something very like the face of our remotest past: a reconstruction of his living appearance, taking into account all the remains that have been discovered, might reveal something like a very short-faced pygmy chimpanzee.

By the 1920s a range of fossil men had been discovered which took the story of human evolution back from the fully modern-looking Cromagnon men of the last Ice Age, through their immediate and much-maligned precursors, the Neanderthal men and the older and even more primitive-looking Java men. Then came such singularities as Piltdown Man, with his plausible conjunction of large cranium and ape-like jaw which seemed so perfectly to confirm the prevailing nineteenth-century view that the progress of humanity had been powered by the enlargement of his brain. Java Man and Piltdown Man were already out of step with each other, but in the prevailing scarcity of human fossils at the time this inconsistency was less noticeable than it was to become later on before Piltdown Man's disgrace.

In 1924 a wholly unexpected candidate for the rôle of remote human ancestor came to light in a limestone quarry in South Africa, and met with no very enthusiastic welcome. An Australian-born anatomist named Dart identified, in a box-full of limestone fragments and fossilised bones from a quarry near Taung, a small skull as belonging to a new family of primates intermediate between ape and man. In fact, he considered that he had found the distant ancestor of the human line. Dart maintained that his find, though the skull of a juvenile (it is often called the 'Taung Baby'), was too large to belong to even an adult chimpanzee: it showed, moreover, no hint of brow-ridges and its newly-erupted permanent molar teeth were more human than ape-like. Developing his inspiration that he had stumbled across man's earliest known ancestor, Dart even conjectured that the forward position of the foramen magnum, the large hole that lets the spine communicate with the brain, indicated an upright walking posture with head balanced on the vertebral column in the human way. Dart called his creature Australopithecus, the Southern Ape, perhaps consciously or unconsciously drawing attention to it as an Australian's Ape at the same time! His claims, more in the nature of bold guesses than scientifically substantiated theories, were not well received by the anthropological establishment, who were inclined to write off the Taung Baby as just another previously unknown fossil ape with no significantly human traits. The eminent anatomist Sir Arthur Keith, who much later espoused Australopithecus whole-heartedly, told the journalists who pestered him about this latest 'Missing Link' after the announcement of Dart's discovery: 'We have a rumour of this kind three or four times a year.' One can only say that we still do! At the time, the opinion of Keith and others that the Australopithecus skull placed him among the apes, made the recognition of his very human leg and wrist bones difficult for most workers. Dart did not abandon his baby but, sadly – for much more evidence from the same source must have been destroyed as quarrying proceeded – he did not for many years pursue the baby's relations and look for more remains of Australopithecus.

The search was continued in 1936 by a 69-year-old one-time Scottish general practitioner and sometime professor of geology named Broom, resident in South Africa since 1897. Investigating limeworks at a place called Sterkfontein, Broom began to find not just skull and teeth fragments of Australopithecus, but remains like ankle bones whose obvious human affinities confirmed his acceptance of all Dart's claims. In 1938 Broom first found the remains of a related creature of heavier build whom he considered sufficiently different from the original Australopithecus finds to dub with another, now obsolete, name. He placed the two sorts of man-ape in a family he called the Australopithecines, a term still useful today. Currently the two creatures are called Australopithecus africanus (the more slightly built form) and Australopithecus robustus.

Immediately after the Second World War, Broom published, at the age of 79, a detailed account of all the Australopithecine finds, and the majority of anthropologists came round to many of the views so cannily canvassed by Dart twenty years previously, concluding that the Australopithecines, whilst not belonging to the genus Homo, stood firmly at the start of the hominid line. Dart returned to the quest for Australopithecus remains and Broom went on to find evidence that clinched the association in the Australopithecines of an 'ape-like' cranium with a man-like jaw and teeth that had upset the expectations of the established anthropologists. Furthermore, with a nearly complete Australopithecine vertebral column, pelvis and hip joint, he demonstrated that the upright walking gait of these creatures was a certain fact and not just a brilliant conjecture of Dart's. A new picture of human evolution was becoming firmly established: one in which the early adoption of erect bipedalism had much preceded the enlargement of the hominid brain.

The Australopithecines were clearly seen to represent an early strain of the hominids, whether or not any of them might constitute the direct ancestors of the genus Homo and modern man. There was a problem about their dates: the geology of the limestone cave deposits in which they were turned up was complicated and there was no method available by which absolute dates could be assigned to them. No progress was made with the dating problem in the 1950s and 60s, and then the limelight was switched from South Africa to East Africa, where creatures very like the South African Australopithecines began to be discovered in 1959.

Olduvai Gorge in Tanzania became famous in that year for the find of a hominid skull whose massive proportions and powerful back teeth led to its nickname 'Nutcracker Man'. It was found by an expedition led by Louis Leakey, who had discovered the first stone tools in the gorge back in 1931. Further tools were found subsequently and scraps of fossilised hominid remains, including in 1955 a tooth thought to belong to an Australopithecine. The skull found by Leakey's wife in 1959 was clearly of an Australopithecine character, but it was found in association with chipped pebble tools – the first time such an association had been conclusively observed. A method of rock-dating by means of potassium-argon content had been newly developed at the time and yielded a date of one and three-quarter million years ago for the level in which the hominid remains were found: the first absolute date assigned to a fossil of Australopithecine affinity, and surprisingly earlier than the guessed dates then attributed to the South African finds. 'Nutcracker Man' possessed huge cheek teeth, a massively fortified skull and a long face too heavy-set to accommodate even the Australopithecus robustus skulls known from South Africa, which are the closest parallels in other respects. Since the original 'Nutcracker Man' find, other similar remains have been discovered in East Africa, and there is no doubt that these creatures are Australopithecines, though perhaps belonging to another species than robustus, as well as africanus. It now seems likely, however, that 'Nutcracker Man' may very well not have made the pebble tools with which his remains were associated. For, in the 1960s, further hominid remains came to light at the same site which were slighter-built and more 'gracile', as anthropologists say, to the point of a close resemblance to the Australopithecus africanus remains from South Africa – but with bigger brain-cases, capable of containing 680 cc of brain as against the average of 450 cc of the africanus specimens. Creatures of this sort, whose remains have been found spanning a period of some three-quarters of a million years, were probably the makers of the pebble-tools; their surviving hand bones have been judged fully capable of the work. They have been assigned the species or subspecies name of habilis, the 'handy' man: some researchers seek to place them on the human line and classify them as Homo habilis, but their affinities with Australopithecus are so marked that others prefer to regard them as a late, perhaps subspecies, form of Australopithecus africanus.

An even bigger-brained member of the Australopithecus family was discovered by Richard, son of Louis, Leakey in 1972 by Lake Turkana in Kenya. Seemingly of about 750 cc, the large-brained ER 1470 skull again resembles Australopithecus africanus very closely and dates to about one and three-quarter million years ago; in 1969 an Australopithecus robustus skull was discovered at the same level, pointing to a similar association of africanus-types with tools and robustus-types as was discovered at Olduvai.

The age of the Australopithecines has recently been pushed much further back than the dates of Olduvai and East Turkana: from Ethiopia have come the remains of many individuals dated to about three million years ago, the most famous of whom is Lucy, a 3 ft 6 in (approximately 1 m) tall female whose skeleton was found in a fairly complete form, though the skull is very fragmentary. At the same time, Mary, wife of the late Louis Leakey and mother of Richard, found Australopithecine remains closely similar to the Ethiopian ones in the Laetoli fossil beds near Olduvai Gorge that are dated to three and a half million years ago. On the grounds of more primitive teeth and other features (including suggestions of a more arboreal life), the Ethiopian and Laetoli finds have been classified by some workers as a new species of Australopithecus, afarensis (who was perhaps the ancestor of both africanus and robustus), but a contrary opinion puts them all into Australopithecus africanus. The oldest fragments of an A. africanus jaw-bone have come from a place called Lothagam, south of Lake Turkana in Kenya, dated to five or six million years ago. Since the latest, habilis, form of Australopithecus reached down to about one million years ago, it is clear that we are dealing with a long-lasting hominid among whose surviving remains we may reasonably expect a fair degree of variation in size and form.

There is a broad division of the genus Australopithecus, into two species, the 'gracile' africanus and the heavily-built robustus. A. robustus seems to have been adapted to the chomping of a fibrous vegetarian diet with his huge cheek teeth and massive musculature of cranium and jaw. It is reasonable to dismiss him from the direct human line of descent as a side-branch on the tree of human evolution, probably evolving out of the gracile form towards extinction after a long period of co-existence with africanus. A. africanus exhibits a considerable range of body-size and brain-size, with the East African specimens including larger forms than the South African ones do. Very likely the brain-size is often related very directly to the body-size of africanus individuals, but it seems equally true that some strain or strains among the africanus populations tended over the millions of years of their existence towards relative and absolute increase of brain-size. After they had been walking about on two legs with hands free for at least one million years, some of the africanus groups took to tool-making, producing the simple pebble tools sometimes found in association with them: a step which marks the beginning of human culture. And in the last million years of their career, some of the africanus Australopithecines – in the shape of the habilis form – began to develop bigger brains which foreshadowed the cranial capacities of the next main stage in human evolution, known to us in the forms of Java Man, Pekin Man and the like.

Many reconstructions of the general appearance of the Australopithecines have been made: of course we do not know anything about the skin colour of these creatures and we can only guess at their degree of hairiness. Perhaps, like the present day native inhabitants of tropical Africa, they were dark-skinned and probably they were as hairy as the chimpanzees and gorillas who constitute their not-too-distant relatives. Australopithecus was short, from 3 ft (0.9 m) to something over 4 ft (1.3 m) according to sex and local variation; he stood upright and walked on his two legs as well as any representative of Homo sapiens today; he had a small brain ranging from a third to a half of the present day average enclosed in a small skull behind marked brow-ridges and a long and prominent face, with no chin to his lower jaw.

Skilled reconstructions have been built up on casts of Australopithecine skulls to show the possible living appearance of these remote human ancestors. Of course, such work is open to doubt and error and in the past some rather grotesque reconstructions have been achieved in this way. But the products of one particular worker deserve regard in the light of his spectacular forensic successes. The Russian archaeologist and ethnographer, the late Professor Mikhail Gerasimov, on several occasions aided the police in identifying the remains of missing persons by reconstructing the appearance in life of skulls found in the course of police duties. His reconstructions bear such close resemblance to the photographs of the missing persons being sought, that we can place a fair degree of confidence in the reconstruction work he did upon such historical entities as Schiller and Ivan the Terrible, and such prehistoric ones as Cromagnon, Neanderthal and Pekin Man. Gerasimov worked on some of the Australopithecine remains too, including the Taung Baby and a female adult from Sterkfontein. For the Taung infant, Gerasimov reconstructed the missing parts of the skull and corrected the deformation of the fossil's appearance (due to squashing before fossilisation). He employed for the reconstruction of the soft parts, measurements from three to five year-old chimps and human children; the ears he modelled on those of a baby chimpanzee and he added hair according to chimpanzee characteristics. For the adult female A. africanus, Gerasimov had to add a lower jaw based upon the teeth disposition of the upper jaw and the indications of jaw musculature on the skull and upon the evidence of other Australopithecine finds. Again he used measurements of the thickness of chimpanzee soft facial parts and furnished his reconstruction with the deep-sunk eyes of the great apes and their thin, flattened lip form; the nose he made very small and not wide, in keeping with the original specimen's narrow and almost flat nasal bones and small nasal aperture.

The reconstructions of Gerasimov allow us to peer a little uncertainly into the oldest known human (or at least nearly human) face of the past – there is a place in Africa which comes tantalisingly close to revealing more of the living form of Australopithecus. At the Laetoli fossil beds in Tanzania, a very rare geological circumstance, brought about when an ancient volcano put down layers of ash at the same time as intermittent rainshowers were falling, has preserved the footprints of two or three Australopithecines as they crossed the still-wet ash, before chemical action and drying-out cemented the layer about 3.6 million years ago. Two tracks (among those of birds and mammals) suggest that a larger (about 4 ft 6 in (1.5 m) individual and a smaller one (under 4 ft (1.2 m)) walked perhaps together – but not abreast, for the tracks are too close – across the ash surface and then another large individual walked in the footsteps of the first! The fully upright walking gait of Australopithecus is perfectly confirmed by these tracks and the feet forms are fully human, with proper arches and without the splayed toes of the apes. What a pity that one of the Australopithecines did not fall face-downwards in the drying tuff!

Chapter Three

The False Face of the Past

Piltdown Man, about AD 1910

Sometime in the early years of the present century, some person or persons unknown took some fragments of a human skull that was no more than five hundred years old and set about staining and altering their chemical composition. And the same person or persons unknown also took the jaw of an orang-utan, about the same age as the human skull, and after breaking it up and removing all the distinctively ape-like details – like the knob that articulates the jaw with the skull – and filing down the characteristically ape-like teeth, likewise stained the end-product along with the skull fragments. The hoaxer or hoaxers also assembled a collection of very old animal fossils and fabricated some fake stone tools. What were the forger or forgers of these bits and pieces trying to make? Of course, they were trying to make 'The Missing Link'! And to make it in keeping with the ideas of human evolution that were current at the turn of the century, which required that our remote human ancestor should flaunt, however ape-like the rest of him was below his atlas vertebra, a large human brain.

The Missing Link that the turn-of-the-century forger or forgers came up with became the most famous prehistoric man in Britain. It made great claims to be the most important fossil of early man in the whole world. They even named the local pub after him, near by Piltdown Common in the parish of Fletching in Sussex, where his remains were supposed to have been found over a period between about 1908–1913.

The background to the search for the Missing Link which Piltdown Man was meant to supply was the gradual dawning in the nineteenth century of the notion of the great antiquity of man and his relationship with the rest of creation. The orthodox view until well into the nineteenth century, based upon a literal reading of the Bible, was that the world had been created in 4004 BC, looking just as it does today with all its geological features and myriad kinds of living things exactly as we know them. Something of the great geological age of the Earth had begun to emerge at the start of the nineteenth century – men began to realise that the Earth had passed through many geological epochs very different from the present one and must be much older than the Biblical calculations allowed. Furthermore, the strata of these remote epochs often contained the fossil remains of creatures no longer existing in the world.

One great authority, Georges Cuvier in France, put forward an account of the geological discoveries that appealed to many: he said that there had been many successive creations, of which the Biblical account dealt only with the latest. There had been many stages of the world prior to the present one and in some of them, the extinct and fantastic animals whose remains the geologists were finding, had lived. But Cuvier was adamant about one thing – human beings had not existed in any one of these previous creations. But at Hoxne in Suffolk, in 1797, a country gentleman named John Frere (interestingly enough, an ancestor of Mary Leakey) found some flint implements, which he rightly considered to be of human manufacture, at a depth of 12 ft (4 m) and in association with the bones of extinct animals. But his sensible observations went unheeded: 'the situation in which these weapons were found may tempt us to refer them to a very remote period indeed; even beyond that of the present world'.

The man who perhaps did most by his persistent searches, in the face of much disbelief from the savants of his day, to establish that the stone tools he and others were finding in the gravels of the ancient water-courses of rivers like the English Thames and French Somme really were the handiwork of men who had lived in a very remote geological epoch, was a French customs officer named Boucher de Perthes. His abundant finds gradually convinced the scientific community of the truth of the claim that men had existed for a much greater period of time than the orthodox and Bible-based account allowed.

The work of two Englishmen finally stabilised the picture of man's antiquity. Charles Lyell vastly extended the geological estimate of the age of the Earth by insisting that all the long record of the rocks with their successive changes, was to be accounted for only by the slow working of the ordinary and natural processes (of constant erosion and occasional vulcanism, for example) that we observe in action in the world today. And Charles Darwin put forward the first convincing picture of how, just like the long slow changes in the physical state of the world, so too the whole of the world of living things could have slowly evolved always from lowlier forms into the vast differentiated natural history of the fossil record. Though Darwin did not say so at first, mankind was clearly implicated in this chain of slow evolution: human beings had evolved out of more primitive and therefore somewhat ape-like ancestors, the apes being clearly our closest relatives in the natural world. Darwin himself and the bolder of his followers went on to speculate about the evolution of man. The German anthropologist Ernst Haeckel produced an imaginative reconstruction of what a form intermediate between man and ape might look like, half-turned away from us in the artist's impression out of shyness about his plausibility perhaps, and brandishing a small tree. But, people asked themselves in the latter part of the nineteenth century, if Lyell is right and the world is very old, and if de Perthes is right and the earliest tools of man are to be found in abundance in very ancient river gravels, and if Darwin is right and man is descended from some more ape-like predecessor, where are the bones of our distant ancestors?

Ironically, in 1856, three years before the publication of Darwin's *The Origin of Species*, a skull-cap had come to light in Germany which belonged to an earlier type of man – the famous Neanderthal Man. But he was not at the time recognised for what he was and variously written off as a cripple or a pathological idiot. In time more and more of his remains were discovered in other parts of Europe and the real existence of the Neanderthal type of man was recognised, but by then it was obvious – despite the misleading reconstructions that made him far too bent and shambling – that he was too young in date geologically and too like us in his humanity to qualify as the remote ancestor of man who had made the crude stone tools found in the very old geological deposits. Boucher de Perthes himself, who had found so many of those stone tools, thought that he had come across the ancient jaw of one of the makers of the hand-axes, but unfortunately three Englishmen were able to prove that he had been the innocent victim of a hoax, perpetuated for reward by some of the men he paid to find things for him. Part of the proof of this deception involved chemical tests upon the jaw which showed that it was not an ancient fossil at all, but a modern bone.

What was at the time genuinely the oldest and most primitive form of man discovered, was unearthed in Java in the 1980s by a Dutchman, Eugene Dubois. Unfortunately, at that time neither the great geological age nor the true evolutionary character of these remains was recognised. Even their discoverer, who initially assigned to them their proper place as a very ancient and primitive form of man, came to doubt them, inclined to view them only as remains of some extinct giant gibbon, and locked them away from all inspection.

1907 saw the find of a human fossil that probably had more to do with the cooking-up of the Piltdown Hoax than any other discoveries. It was an ancient human jaw from a place called Mauer, near Heidelberg in Germany. It dates from about the same time as the Java remains and represents roughly the same stage of human evolution. But whereas the Java remains were locked away and not widely known, the Mauer jaw was treated to a lot of publicity. Easily its most noticeable feature was that it was very massively made.

This was the state of affairs with the study of human evolution in the years before Piltdown Man began to make his entrance: there was no really convincing and recognised candidate for the rôle of remote human ancestor, certainly no very complete one – only the massive solitary jaw from Mauer, announced to the world in 1908.

The story of the discovery of Piltdown Man is not an easy one to piece together. It has to be said at the beginning that, whether or not the principals involved who gave their own accounts of the find were themselves involved in the fraud, they certainly did not keep clear records, or make proper plans and sketches, or take photographs of their work. They are extremely vague as to dates, sometimes even where the years of the finds are concerned.

It was a lawyer and amateur antiquarian of Uckfield in Sussex who made the first discoveries. One of his professional duties brought him about every four years the short distance from Uckfield to the Piltdown area, where he was Steward of the Manor of Barkham. Charles Dawson seems first to have entertained the idea of finding fossils in gravel diggings undertaken for road-mending material around Piltdown Common in 1904 and 1907. In 1908, it seems he was given a 'small portion of an unusually thick human parietal bone' – a fragment of the side-wall of a human skull – by one of the workmen he had previously alerted to keep an eye open for fossils. He searched for more where the piece of skull had come from but 'in fact the bed seemed to be quite unfossiliferous'. Other versions of the story have it that the workmen found and smashed what they took to be a coconut, only keeping one piece to show to Dawson. Dawson claimed that it was not until the autumn of 1911 that he picked up, on another visit to the spot, another and larger piece of the same thick skull which he and the workmen had missed a few years before. One of those who later vouched for having seen a piece or pieces of the skull at around this time, was the young friend of Dawson's, Teilhard de Chardin, who was at the Jesuit seminary in Hastings in 1909. (He became famous in the 1950s, after a life-time as a palaeontologist, as the writer of very obscure but voguish works reconciling science with religion.) Dawson also informed Sir Arthur Smith Woodward at the British Museum of his skull find and sent him a fossil tooth of an ancient form of hippopotamus from the same site, which seemed to confirm Dawson's conviction of the great age of the deposit and the skull. Later, Dawson took the skull fragments to London to show to Smith Woodward, with the dramatic challenge: 'How's that for Heidelberg!'. In addition to the skull and the hippo tooth, Dawson also revealed more ancient animal fossils and some very primitive flint tools to Smith Woodward: and the hunt was now on for the Sussex Missing Link.

When Dawson, Smith Woodward, Teilhard and a local workman

began digging at the Piltdown pit in June 1912, new finds quickly came to light. Smith Woodward personally recovered some more fragments of the thick human skull and Teilhard found the tooth of an extinct form of elephant. More flints, assumed to be tools, came to light, Teilhard again finding one of the more spectacular items. And one day in that June of 1912 there came flying out of the gravel before them, when Dawson and Smith Woodward were digging without the benefit of Teilhard's assistance, an astonishing piece of jaw-bone to go with the skull fragments already collected: its colour, a deep red like that of the gravel in which everything was found, perfectly matched the skull fragments.

In December 1912, the fruits of the Piltdown gravel-pit to date were gloriously announced to the Geological Society of London. In discussing the significance of the finds, Smith Woodward took the view that the skull pieces and the jaw all came from the same creature. In fact, the cranium was not really to be distinguished from a modern one, except on on the grounds of its unusual thickness – although one famous scholar pronounced the brain cast to be the most primitive he had ever seen – while the jaw was very ape-like indeed, except for its teeth which Smith Woodward said were nothing like an ape's and much more like a man's. Unfortunately, the give-away parts of the jaw which would have settled whether it articulated onto a hominid or a pongid skull were broken off. At all events, if you took Smith Woodward's line that the skull and the jaw came from the same creature, then you had before you a sort of human being, blessed with a nobly human brain but still lagging behind a bit in the refinement of his jaw from ape-like origins. Though there were arguments about the precise reconstruction of the pieces, this opinion tended to carry the day: after all, this conjunction of human brain with apeish jaw was just what everyone had been looking for and expecting to turn up in the way of a remote forefather of the human race. And at Piltdown, here were the crudely worked flints of the human ancestor, too, and the bones of some of the now extinct animals with which he had shared his ancient world.

But there were a few present at the Geological Society meeting in 1912, and there were to be more in the years to come, who took another view of the matter. A professor of anatomy and a dentist, to their eternal credit, said simply that the skull and the jaw could not be of the same creature – the skull was too human and the jaw was too ape-like, they must have accidentally come together in the gravels. What was more, according to the dentist whose views were not properly respected, the state of the wear on the teeth was very odd: you could not arrive at wear like that, worn all flat, on teeth that appeared only just to have erupted. His shrewd observations were not followed up, and Eoanthropus dawsoni was christened that day – Dawson's Dawn Man – complete with a zoological classification that we now know he had no claim to, the non-existent species of a non-existent genus.

More finds followed. In 1913, Teilhard de Chardin was lucky enough to find a canine tooth, absent in the jaw as found, in a spread of gravel where Dawson and Smith Woodward had missed it. It fitted predictions made about its likely shape and size at the 1912 meeting to a 'T'. A unique and preposterous whittled implement made from a piece of the leg bone of an elephant showed up under a hedge, but was judged by Dawson and Smith Woodward to have come somehow from the deep gravels and so belong with the rest of the stuff: it is nothing more nor less than a blatant cricket bat carved in bone.

Most spectacularly, in 1917, Smith Woodward was able to announce the finding of the remains of a second Piltdown Man. Dawson had died in 1916, but before he died he had told his old friend of his fresh success in finding the remains of Piltdown II at another site a couple of miles from Piltdown in 1915. Smith Woodward never learned the exact spot from his ailing friend and, though he visited the place with Dawson before going home to France, Teilhard could not afterwards remember the exact circumstances. Two pieces of skull with a molar tooth and the tooth of a rhinoceros made up this second find. For even a number of the former doubting Thomases, this second find settled it. You could not expect that a couple of prehistoric men and a couple of prehistoric apes had accidentally come together in illicit union in two places some miles apart, and left their bones promiscuously entangled to confuse the anthropologists of the twentieth century. These must be the remains of two similar ape-men, two genuine Missing Links, who had lived in Sussex, as the animal remains from both sites indicated, an extremely long time ago. So, though doubts persisted in some scholars' minds (particularly in the United States), as to the true association of the Piltdown finds, this bogus and composite monster took his uneasy place in the textbooks. The popular press took him up with enthusiasm and, of course, 'scientific' reconstructions of his appearance in life were produced on the basis of the anthropologists' attempts to put his fragmentary bones together into one whole skull: attempts made very difficult for them by the fact, which they did not recognise, that the bones did not come from the same individual, nor even from the same zoological family of primates, being made up as they were of the cranial bones of a recent hominid and the interfered-with mandible of a pongid – a modern man and an orang-utan. When we try to meet the shifty eyes of Piltdown Man, as reconstructed by Smith Woodward and the artists of the popular journals of the 1920s, we are looking into the False Face of the Past, the impossible physiognomy of the 'Ape-Man who Never Was'.

In fact, throughout the '30s and '40s, Piltdown Man's place in the textbooks and in the consideration of the scholars became increasingly uneasy. His face just did not fit in the company of the new finds and discoveries that were coming along. These new finds began to build up a particular picture of human evolution that did not square with Piltdown at all. The remains of the Australopithecines started to appear in South Africa, evidently very old, and they disclosed a sort of creature whose jaw and teeth were more human in form than Piltdown's, but whose brain was by contrast tiny and quite inferior to the human status of the Piltdown brain. The old Java finds were rediscovered from their excavator's concealment of them and new finds were made in Java. This time the Java men were appreciated for what they are. Related human remains came to light in China, and the stage of human evolution represented by Java and Pekin Man was seen to follow naturally enough upon the Australopithecine stage. With Java and Pekin Man, the rest of the skeleton below the head had achieved a more or less fully human size and form: only the heavy face and still inferior brain-size attested to the more primitive standing of these men beside modern human beings. They made entirely plausible ancestors for Neanderthal men and then ourselves.

Piltdown Man was the odd man out. With his nearly ape's jaw and his full-size brain, he did not fit into the emerging picture of human evolution at all. To make things much worse, new geological datings of the gravels in which he was alleged to have been found meant that he could not be more than 50,000 years old at the most. An ape-man, or even an ape and a man, running around in Sussex at a time when Homo sapiens had already appeared and there were certainly no apes in Europe, was a clear impossibility. Anthropologists did not like to talk about Piltdown Man: he was an embarrassment.

In 1953, Piltdown Man was ignominiously retired from the world of science: the whole assemblage of remains constituting Piltdown Man was proved to be fraudulent. Fluorine-content tests on the skull and jaw fragments showed them to be of different ages, though neither was very old: their interior characteristics were quite dissimilar. Both had been stained artificially, but the porous cranial fragments were stained through while the jaw was only superficially stained. The jaw had been subjected to breakage to remove its most obviously ape characteristics, and the teeth in the jaw had been filed with a metal file to give them a more human appearance. The canine tooth was revealed under X-ray to have only recently erupted, yet it was worn all over: it had actually been painted with brown paint! The stone tools and some of the animal bones associated with Piltdown Man had also been artificially stained – the cricket bat was shown to have been an already fossilised piece of bone when it was whittled with a metal knife. Piltdown Man was, in short, a fake from first to last, and people naturally began to ponder who could have done it. Scores of candidates have been proposed, many quite implausibly.

It seems likely that the whole Piltdown affair began with Dawson's amateurishly uncritical, but initially innocent self-persuasion that he had found the Oldest Man in the World. It would not be out of keeping with Dawson's known character and other exploits to consider that, in order to convince Smith Woodward (not, it turned out, a difficult task) of the antiquity and importance of his find, Dawson faked and salted in the gravels the fantastic jaw. Teilhard de Chardin may have guessed what had happened and himself puckishly enlivened the proceedings with the canine tooth (in a way that must have dumbfounded Dawson, who had no option but to press on and hope for the best). Someone at the British Museum, perhaps a group of them, who disliked Smith Woodward and despised his abilities, decided to take the matter further, contributing such manifest absurdities as the cricket bat in the expectation of rapidly bringing the whole nonsense to a comical, and for Smith Woodward embarrassing, end. When their efforts went unrewarded and the scientific community, with honourable exceptions, continued to swallow Piltdown Man, they gave up the unequal struggle and Dawson was even able, almost from beyond the grave, to foist a second Piltdown Man upon the world.

The episode affords its lessons: it reminds us that even eminent scientists are liable at times to see what they expect to see or want to see when their preconceptions are engaged, and miss what doesn't suit them. More flatteringly for the present-day status of the science of anthropology, the Piltdown story highlights the enormous advances that have been made in scientific testing methods and science-based means of dating in this century, advances which make the perpetration of similar frauds difficult to the point of practical impossibility today.

Chapter Four

The First Men

1,000,000–200,000 Years Ago

The German palaeontologist Ernst Haeckel was an early supporter of Darwin, who turned his attention to the question of human evolution. Before any fossil remains that might belong to a really early stage of human evolution had been found, he speculated about the sort of creature that would fit the bill as remote ancestor linking the human line with the apes. He even invented a full scientific name for this absent creature: Pithecanthropus alalus. Pithecanthropus means ape-man and alalus means speechless, since Haeckel thought that so early an ancestor of man would have been incapable of speech. Inspired by Haeckel's conjectures, artists produced impressions of what Pithecanthropus might have looked like. When we regard the faces of these reconstructions, we are looking not into the False Face of the Past, as we do with Piltdown Man, but at the Speculative Face of the Past which, with allowance made for the growth of knowledge with the progress of anthropology, is a reasonable enough thing to be invited to do.

In the 1880s Haeckel's speculations fired the imaginations of many people to want to search for the bones that would substantiate them. One young man who was prepared to go to great lengths to look for fossil men was a Dutchman named Eugene Dubois. As he was Dutch, his thoughts turned naturally enough to the Dutch East Indies, which were not only accessible to him, but fulfilled the Darwinian requirements of being tropical and affording living examples of man's closest relatives in nature, the apes. Darwin himself, it is worth noting in the light of subsequent discoveries, favoured Africa as the cradle of human evolution.

Dubois arranged himself a job as army surgeon in the East Indies and was actively looking for fossils there by 1890. In 1891 he began work near the village of Trinil on the Solo River in Java and found first teeth and then cranial remains with astonishingly rapid good fortune. His skull pieces struck him immediately as those of some primate bigger than an ape and smaller than a man – just what he was looking for. In 1892 at the same place, in the same geological level and only 45 ft (14 m) in distance away from the site of his original finds he discovered a fossil femur (thigh-bone) that surely belonged to the same sort of creature, if not actually the same individual, as the skull. In 1894 Dubois published his finds as Pithecanthropus erectus, borrowing Haeckel's genus name but emphasising the upright walking, that he was sure the very human-looking femur indicated, over the putative speechlessness of his ape-man. The reception accorded to P. erectus by the scientific community was mixed: some thought him an ape pure and simple, some thought him human (perhaps an idiot) and some thought him intermediate between apes and men. There was naturally some argument about whether the very human femur and the not-so-human skull belonged together, a notion which at that time was not capable of proof, despite the finds having come in close proximity from the same level in the same state of geological preservation. Interestingly, it was fluorine testing that later showed them to be of the same age. When Dubois' own conviction that he had found the intermediate form between ape and man ancestral to the human line was not universally espoused, he retreated with his finds to Haarlem in Holland and, for a period, appears to have kept them locked in a box under the boards of his dining-room floor! It was more than 20 years before the bones went to Leiden Museum and Dubois himself changed his mind about what he had found more than once, settling in the end for the opinion that the fossils only represented the remains of an extinct giant gibbon, in contradiction of his earlier and better thoughts.

In the late 1930s more discoveries were made in Java; von Koenigswald found fragments of jaw and skull at a place called Sangiran. By the outbreak of the Second World War, a respectable population of Java Man finds had been unearthed, which meant that the standing of Java Man need not rest on a single individual. The remains belong to a range of dates from about one million years ago to about 700,000 years. The femur suggests that these fellows stood about 5 ft 6 in (1.7 m) tall, but with a slender frame. Their cranial capacities range from a low of 775 cc, overlapping with the latest and most evolved Australopithecines, up to perhaps 975 cc, at about the lowest limit of modern human beings. Java Man's post-cranial skeleton was very like a modern man's, but from the neck upwards he differed considerably. His neck must have been very short to judge by the muscle attachment areas on his skull, and he would have needed that short neck to carry his thick and heavy skull, with its huge face and jaw and very large teeth. But his teeth, though large, are much more modern-looking than those of Australopithecus. The face of this man of the past was rather projecting: his skull was widest just above the ears and sloping rapidly in to a ridge along the top of his head, with really no forehead at all behind the bony and pronounced brow-ridges.

In another part of the Far East, there came to light in the late '20s and '30s another closely related form of early man: the famous Pekin Man, whose remains were to meet an unhappy end in the Second World War. From a site called Choukoutien about 25 miles (40 km) south-west of modern Beijing, there came the first teeth and fragments of jaw and skull, and then in 1929 the first more or less complete cranium, found by Pei. A German anatomist named Weidenreich took up the case of Pekin Man and in 1936 parts of seven skulls were discovered. All in all, the remains of more than 40 individuals were found, as well as the stone tools they fashioned, and evidence of fire and cannibalism. But after 1937 the Japanese, who were expanding into northern China, started to show an interest in acquiring the bones of early man found in territories they regarded as their own – Weidenreich left for New York in 1941 taking with him some very detailed plaster casts of the Pekin Man remains. The authorities of the Geological Survey of China decided that the fossils themselves should follow Weidenreich for safe-keeping in America. Their boxes were placed on a train on 5 December 1941, but the American ship on which they were to be loaded was captured by the Japanese in port: we know that the ammunition carried on the same train as the boxes fell into Japanese hands, but it seems sadly likely that the boxes of fossils went to the bottom when the ordinary Japanese soldiers who ransacked the train came upon them.

Fortunately Weidenreich's Chinese-made casts and his own detailed studies have proved more than adequate for the scientific investigation of Pekin Man. He turns out to have been very closely similar to Java Man, only bigger-brained. Weidenreich himself thought Java and Pekin Man no more different from each other than modern races differ in the world today, but in view of the brain-size differences and the difference in geological age, it seems reasonable to regard Pekin Man as a later and more evolved form of the same species as Java Man. Pekin Man seems to have flourished around 300,000 years ago, many hundreds of thousands of years after the youngest Java Man of about 700,000 years. Both of these ancient forms of men are now classified unambiguously as belonging to the genus Homo (the genus Pithecanthropus is obsolete since Java and Pekin Man clearly belong to the same genus as ourselves and do not constitute 'ape-men'), but they are assigned not to the species sapiens like ourselves and Neanderthal Man, but to the different species erectus. In both Java and Pekin Man the skeleton below the neck is virtually indistinguishable from modern man's, but both possess thick-walled and beetle-browed skulls. However, Pekin Man averages a cranial capacity of 1075 cc, as against Java Man's 850 cc. Pekin Man's skull is higher-sided and he displays the beginning of a forehead bulge so noticeably lacking in Java Man. Pekin Man's teeth are smaller and consequently his mouth has a more human-looking configuration, with a shorter jaw and, for the first time, a hint of the chin that is a distinctive feature of Homo sapiens.

Since the Second World War the original remains of Pekin Man have remained absent, but some more fragments have been excavated at Choukoutien. An older and smaller-brained type has been discovered at another site in China, perhaps contemporary with Java Man, but perhaps older at over one million years ago.

Other remains of the same general type and classified as Homo erectus have come to light in North Africa and Hungary and, of course, the Heidelberg Man jaw found at Mauer in 1907 must belong to the same stage of human evolution. The Mauer jaw, though very massive, displays smaller and more modern teeth than the Pekin Man, and the Hungarian fragment, at perhaps about the same date as Pekin Man, points to a larger brain of about 1300 cc, well into the modern range. From Olduvai Gorge in Tanzania has come a Homo erectus with an early date of one million years ago and a rather large brain for his time at 1000 cc, while from a site in Israel, which lies on what may have been a corridor between Africa and Asia, come some possibly erectus remains that may be nearly two million years old! Homo erectus had a long run and, even after more modern forms of men had evolved in some parts of the world, in other places erectus traits lingered on: in Java, with the Ngandong skulls of about 100,000 years ago, in Australia, at a very late date of only a few thousand years BC, and in Africa, where the skull from Broken Hill in Zambia shows clear erectus features in his massive face and brow-ridges at about 130,000 years ago.

Of course, Homo erectus did not evolve into Homo sapiens everywhere in the world at the same time, and there were places where erectus traits remained noticeable even after the threshold to Homo sapiens and even Homo sapiens sapiens had long been crossed: Australia seems to be such a place. It has been proposed that the broad racial divisions of humanity today may be in part very old and go back to erectus days: from Choukoutien come skulls much later than Pekin Man which do seem to maintain some of Pekin Man's characteristics and to foreshadow the distinctive racial features of the Mongoloids. Similar erectus-rooted racial traits have been proposed for the Australoids, Caucasoids and (though fossil evidence is scanty) the Congoids and Capoids of Africa. On this view of human evolution, mankind has evolved globally in genetically distinct (but not isolated) geographically-anchored populations through the same stages of erectus and sapiens – though not necessarily at exactly the same times and with exactly the same patterns of features. This view of the origin of races conflicts, of course, with the more generally held one which proposes that all the modern diversification of racial types postdates the arrival of the first Homo sapiens sapiens population a few scores of thousands of years ago.

Many attempts have been made to reconstruct the facial features of Homo erectus ever since Dubois first discovered his Java Man. Artists have drawn and painted him, blessing him with varying degrees of hairiness and perhaps sometimes going too far in that direction, though we have no physical evidence to rely on; sculptors have modelled up his features, sometimes avoiding the hair issue altogether by showing him bald, and sometimes adding lanky locks. Gerasimov has turned his hand to a range of erectus finds. He tells us that he has generally assumed soft parts very much on the modern human pattern, especially with his female reconstructions. He thinks that sexual dimorphism, differences of bulk and robustness between males and females, was more marked in erectus than in modern man, and on a par with that seen among chimpanzees. But with males, too, Gerasimov leaned on the modern human form for the reconstruction of the soft parts of the face, modifying these only where the skull relief was exceptionally marked. Again, Gerasimov's work is to be taken seriously in the light of his successes with police investigations and of his lifetime study of his subject. His reconstructions, and those of other modern workers like Maurice Wilson for the British Museum, agree very well and offer us an opportunity to look with some assurance of plausibility into the faces of our ancestors of a quarter of a million to a million years ago.

Chapter Five

Early Homo Sapiens
250,000–30,000 Years Ago

When the great Swedish naturalist Linnaeus was working upon his classification of living things in the eighteenth century (his is the system which we basically still employ, with its genus-and-species designations like Homo sapiens), he placed man firmly among the animals and in the company of the apes. Linnaeus even considered that his genus Homo might include some other living species besides sapiens, the species name reserved for walking, talking, *knowing* man. He entertained Homo nocturnus or Homo sylvestris orang-outang, on the basis of travellers' accounts of the orang-utans of Borneo and Sumatra – the 'Old Men of the Woods' – whose casual appearance does closely resemble human beings.

Anthropology today recognises two sure species of the genus Homo: erectus and sapiens. There are arguments about whether some among the Australopithecine and related forms should be included in the genus Homo, but erectus and sapiens are the certain species. At one time, Homo sapiens was applied only to the very recent sorts of men who have lived since about 35,000 years ago, excluding Neanderthal Man as a separate species althogether. But now H. sapiens is used in a wider sense to embrace all the forms of man that have succeeded H. erectus.

For the earlier part of the succession, human remains are few and far between. Dating from around a quarter of a million years ago, there are some tantalising scraps which do hint at the evolution of H. erectus towards sapiens. There is a jaw from near Toulouse in France that shows a sort of Heidelberg Man–Neanderthal Man mix of features; from a cave site in the South of France comes a skull which has been claimed to exhibit half erectus and half modern traits; other fragments of the same age range are known from Greece, China and Africa. The two most interesting skulls of this period come from Swanscombe in England and Steinheim in Germany: a statistical study has shown that they are very similar and can be taken together to build up a picture of this 250,000-year-old face of the past, intermediate between erectus and ourselves. Both skulls are female and thick-boned and, to judge by the Steinheim woman, possessed pronounced brow-ridges and heavy faces. The Steinheim woman, with a brain size of 1170 cc, exceeds any known erectus female but not some erectus males; the Swanscombe woman at 1270 cc outdoes erectus specimens of either sex. Neither displays the sloping-sided skull shape of erectus seen from the back; both are rounded towards the crown. Gerasimov's reconstruction of the Steinheim woman shows an amiable if somewhat coarse countenance with very noticeable brow-ridges and sloping forehead.

A skull from a place called Fontéchevade in France, estimated to be about 150,000 years old, marks the attainment of a fully modern brain-size, though the face is lacking to tell us exactly how this specimen looked. A brain fully the equal in sheer size, at least, of modern man's is the boast of the Neanderthals as a whole – indeed the average size of Neanderthal Man's brain is actually greater than modern man's, but perhaps differently distributed and in keeping with the heavy if short build of the Neanderthalers. As a matter of fact, the largest fossil human brain-case ever found belongs to a Neanderthal with 1700 cc who had, in many ways, an interesting and eventful life. He was born about 40,000 years ago in Iraq with a withered right arm which was successfully amputated in life; his teeth wear shows that he used his mouth to make up for the deficiency of his arm; he had at some time been deeply scratched down the side of the face by a cave-bear, and hit over the head (possibly blinded) by the proverbial blunt instrument; but he survived all this to die at home by his own fireside when a piece of the cave roof fell down and flattened him where he stood.

A great number of Neanderthal remains have been discovered in Europe, North Africa, the Middle East, Western Asia and beyond: indeed Neanderthal Man may be regarded as a type representing a stage of human evolution all over the Old World (man had not yet reached the New). The oldest remains date back to a time before the onset of the last fierce phase of the Ice Age at about 70,000 years ago, the youngest may reach down to about 30,000 years before the present. Much has been made in the past of a distinction between the earlier Neanderthalers, with a less markedly Neanderthal look about them, and the later 'classic' Neanderthals, but it might be more reasonable to regard Neanderthal Man as a species that always displayed a wide range of variation in physical appearance, just as modern forms of men around the world do, only perhaps even more so. Gerasimov thinks that there was a greater sexual dimorphism between Neanderthal men and women than there is between today's sexes.

The first Neanderthal remains were found in a Belgian cave in 1829 and were not at the time recognised for what they are. A Neanderthal woman's skull was found in Gibraltar in 1848 but not recognised till 1906. The remains which gave their name to the whole species were discovered in Germany in 1856, three years before the publication of Darwin's *Origin of Species*. They were found in a limestone valley near Düsseldorf in the course of quarrying, and the valley's name has a history colourful enough to bear repeating. Tal, previously spelt Thal, is the German for valley: a seventeenth-century poet baptised Neumann was famous for frequenting this particular valley and the Düsseldorfers took to naming it after him. The poet liked to translate his name – which in English would have been Newman – into Greek as Neander: hence Neanderthal. Skull pieces and bits of the rest of the skeleton made up the original find.

Neanderthal Man was to go on to receive a bad press as his remains turned up elsewhere, but from the start his reception was ominous. An eminent anatomist pronounced him to be a pathological specimen and others relegated him to the role of some forgotten village idiot. More perspicaciously, an Irish authority considered him to be a perfectly normal representative of a previous species of man and Thomas Huxley, Darwin's great supporter, thought him – in keeping with the modern view – to have been a closely related ancestral form of the modern type of man.

The popular image of Neanderthal Man as a stooped and even ape-like shambler, unable to haul himself quite upright and lift his shaggy countenance to regard the heavens, is largely due to the prejudiced reconstructions of diseased specimens made early in this century. The principal instance involved a 40-year-old Neanderthal whose whole spinal column had been deformed by arthritis, whose limbs were bent under him and whose lower jaw had lost almost all its teeth before death.

Everything that we know about Neanderthal Man shows him to have been an accomplished fellow. He possessed a tool-kit far more advanced than that of any of his predecessors; he was a very capable hunter whose quarry included big game; and he employed fire to warm himself and cook by. He lived in caves where they were available, and in Russia, where they were not, he built shelters out of the bones and skulls of the big game he brought down. He wore clothes and pioneered the human race's ability to survive in very cold conditions during the last Ice Age. He developed the first glimmerings we can discern of the 'higher' human attributes, with evidence for a cave-bear cult in some Swiss caves and burials with grave-goods from Western Europe across to Central Asia (in Iraq, flowers were cast on to a Neanderthal grave). Among the Neanderthals, even the first faint suggestions of art appear, in the form of occasional meandering finger lines on cave walls, cup marks carved into a stone covering a burial and, from Hungary, a solitary pebble incised with a cross mark.

What did the innovator of all these very distinctively human traits look like? Neanderthal Man was a very thick-set individual, short and squat with males averaging just over 5 ft (1.5 m) and females just under. The Neanderthals were barrel-chested and heavy-limbed, with a noticeably short fore-arm that must have made a powerful lever. In fact, the Neanderthal skeleton is probably the most distinctive among all representatives of the genus Homo, and it has been suggested that its peculiarities constitute a response to the harsh conditions of the Ice Age. But this explanation seems to leave out the fact that later men have lived in arctic climates just as severe without developing the Neanderthal features. It is fair to say that the skeleton of Homo sapiens neanderthalensis does display subspecies divergences from modern man's, greater than those seen among the living races of Homo sapiens sapiens.

But the most noticeable things about Neanderthal Man were concentrated in his face. This Face of the Past was a very large one beneath heavy brow-ridges that ringed each eye-socket separately, rather than running across the base of the forehead as a bar of bone, as in Homo erectus. Neanderthal Man's nasal aperture was wide and his whole mid-face projected forward over an appropriately long jaw that showed no real chin. Neanderthal Man's face was carried in front of a large but low brain-case with a distinctive backward projection where powerful neck-muscles were attached, and the maximum width of his skull occurred not over his ears, as with most of us, but farther back.

Some of the reconstructions of Neanderthal Man attempted by anthropologists, artists and sculptors have probably made him too hairy (in keeping with a picture of him as an adaptation to the coldest phase of the last glaciation, like the hairy mammoth and the woolly rhinoceros), but many recent efforts do capture his essential humanity. Gerasimov has worked on a number of Neanderthal individuals and his work does reveal something of the range of variation of Homo sapiens neanderthalensis. In particular, he has reconstructed two Neanderthal children that exemplify this range: the nine or ten year-old boy from Teshik Tash, south of Samarcand, who displays a larger and heavier look than the same aged child would today, especially in the face, and the perhaps two year-old infant from Staroselie in the Crimea, who seems to Gerasimov more closely to resemble a modern child.

Now that Homo sapiens neanderthalensis and Homo sapiens sapiens are regarded as two subspecies of the same sapiens species of the genus Homo, the question is naturally asked as to what the precise relationship between them may be. In the past it was fashionable to dismiss Neanderthal Man from the line of descent of modern man altogether, preferring to discern present day human origins in the Swanscombe sort of man (before Swanscombe's close relations with the heavy-faced Steinheim were identified) or among some as yet undiscovered ancestors living in out-of-the-way places.

Another school of thought proposed that only the earlier and putatively less classic Neanderthals were concerned in the direct line of descent of Homo sapiens sapiens, while their classic relatives became extinct after tens of thousands of years of Ice Age life. Desmond Collins, in his book *The Human Revolution*, has recently advanced a biological mechanism by which the Neanderthal population as a whole might have evolved into modern man. 'Neoteny' is a process whereby species become sexually mature and capable of reproduction at a less fully adult stage of development. A well-known example concerns the axolotl, a South American salamander that usually reproduces in a larval state rather than going on to mature into a salamander proper. A number of juvenile skulls of Neanderthal Man have been found (and, indeed, of earlier hominids going back to the Taung Baby) and they all – including the ones with classic features like the Teshik Tash boy – more closely resemble adult H. sapiens sapiens skulls than adult Neanderthal skulls do. For that matter, the skulls of baby chimpanzees do the same. It might be said that Modern Man, even in adulthood, displays numerous child-like or 'neotenous' traits. On the other hand, there is a biological mechanism called hypermorphy, whose effects tend in the opposite direction and the hypermorphic disease called acromegaly can sometimes make modern human beings look rather like Neanderthal Man. Collins proposes that H. sapiens sapiens may simply be the direct descendant of H. sapiens neanderthalensis by means of the rapid operation of neoteny, pointing out that several tens of thousands of years may in any case separate the last of the Neanderthals whose bones we have found, from the first of the fully modern men, commonly called the Cromagnons. The opportunity for neoteny to do its work upon the Neanderthals, who after all had already arrived at a modern sort of brain-size, perhaps lay in the effects of human culture: cooking to render meat more digestible and less in need of chewing and the improved tool-kit that brought food to the mouth, if not the 'table', in small and convenient pieces, may quite simply have done away with the need for big teeth in big and powerfully articulated jaws. The reduction of the jaw and its attachment areas on the cranium meant a reduced face and reduced skull ruggedness all round: the most noticeable areas of difference between Neanderthal and Modern Man. Neoteny may have conferred its real evolutionary benefits, physically manifested in more child-like skulls, in the form of more child-like behaviour – for the playfulness, inventiveness and mental flexibility of man are surely among his most useful attributes.

At Mount Carmel in Palestine there was discovered in the 1930s a series of Neanderthal interments which seems to indicate a population in the course of rapid evolutionary development, manifested in both skulls and post-cranial skeletons, over a period from about 40,000 to 30,000 years ago. Some of them have a very H. sapiens sapiens look about them. From a later date range, 30,000–20,000 years ago, come remains from Czechoslovakia that seem to preserve some ruggedly Neanderthal features in an early population of modern men. The oldest of the Cromagnon remains from France and elsewhere probably date back to about 20,000 years ago, by which time we can be sure that Homo sapiens sapiens was well established as the current species of human being.

But before we move on to consider the appearance and the art of the first representatives of Homo sapiens sapiens, and before we try to trace something of the origins of the modern races of mankind, we may ponder a final question about Neanderthal Man. Is it possible, whether or not he was the direct ancestor of Cromagnon Man, that some of Neanderthal Man's own kind have survived in remote places until more recent times, even the present day? Inevitably the Abominable Snowman or Yeti of Tibet and the Bigfoot of North America, notwithstanding the total lack of physical evidence of their existence, have been identified with Neanderthal Man. Even a few archaeologists have speculated as to whether such creatures as the reported (but not photographed or collected) 'Almasti' of the Far East may not be surviving representatives of the Neanderthal line, with the short legs, hairy pelts and heavy brow-ridges mentioned by some witnesses. The idea is not, of course, totally out of the question – though the descriptions of the alleged Yeti and Bigfoot at least do not seem to accord very well with what we know of Neanderthal Man's likely appearance, whatever may be the truth about the Almasti. And there must have been periods in the past when erectus populations co-existed (at least in time if not closely in space) with sapiens populations, just as various sorts of Australopithecines most certainly co-existed. The world is not, however, so vacant and unexplored a place as it was in the past when all populations of hominid were small and thinly spread. So that while the possibility of the survival of earlier forms of man is not altogether to be dismissed, there is at the same time no real evidence to support it.

This lack of evidence is vividly instanced in the case of a claimed Neanderthal body frozen in a block of ice and put on show from time to time over the past 15 years in America. Displayed in a glass-topped and refrigerated coffin, and allegedly giving off the whiff of decaying flesh, the 'creature' purports to be the mortal remains of a Neanderthaler shot within the last 20 years in some remote arctic refuge. A Belgian researcher of rare and exotic animals, including the Abominable Snowman, is reported to have judged that the 'body' in the ice is indeed that of a Neanderthal Man living only a few years ago before being shot and refrigerated. However, Bernard Heuvelmans did not apparently examine the body personally – which the ice-block effectively prevents. It turns out, however, that the 'body' in the ice was made to order, on behalf of clients who did not say exactly what they were going to do with it, by a Disneyland model-maker in the 1960s who has since died; his widow and son recognise his handiwork with certainty. In fact, he was following an artist's conception of Cromagnon Man when he made his rubber model and the fact that the figure in the ice-block, in so far as it can be made out, resembles neither Cromagnon nor Neanderthal Man, only goes to show that most people have not the slightest idea of what these people looked like.

Chapter Six

Homo Sapiens Sapiens

30,000 Years Ago to Present

The last of the Neanderthals whose remains have come down to us may date to about 40,000 years ago; the first of the Cromagnon men to about 25,000 years ago, if we stick to well-dated finds. Between the 'departure' of the former sort of fossil men and the 'arrival' of the latter, there is a gap of approximately 10,000 years or more. During that time the stone tool-kit associated with the Neanderthals, and called Middle Palaeolithic (Palaeolithic means Old Stone Age, as opposed to the New Stone Age of the First Farmers, the Neolithic), was replaced, at about 35,000 years ago, by the Upper Palaeolithic. The Middle Palaeolithic is characterised by the use of broadish flakes of flint to make scrapers and cutting tools: in the Upper Palaeolithic, the emphasis is upon narrow blades of flint of a more sophisticated and versatile range including knives, awls, chisels, engraving tools. The Upper Palaeolithic took-kit undoubtedly facilitated better food preparation, better woodworking, better manufacture of skin clothes (the evidence of hide shirts and trousers has come from some Russian graves of about 20,000 years ago).

Whenever the bones of Cromagnon men are found in association with flint tools, it is tools of an Upper Palaeolithic kind that accompany them. The hunters who wielded the Upper Palaeolithic range of tools and weapons were sometimes prodigiously successful: at a site in France they killed 100,000 horses over the years and at a site in Czechoslovakia 1000 mammoths. But it is clear that the first Upper Palaeolithic 'industries', as archaeologists call them, evolved out of Middle Palaeolithic prototypes. The final Middle Palaeolithic of France, for example, begins to develop the blade forms seen in greater abundance in the earliest French Upper Palaeolithic and the same process is observed in central Europe. It does not look, in the light of this evidence, as though new populations of men invaded Europe to introduce the Upper Palaeolithic, and so it seems reasonable to conclude that the makers of the final Middle Palaeolithic, presumably Neanderthal men of some sort, were the ancestors of the Cromagnons of the Upper Palaeolithic.

Cromagnon Man is named after a huge rock shelter (now the back wall of a fine hotel) called Cro Magnon (which means in the local patois Big Big!) in the village of Les Eyzies in the Dordogne area of France. The Cromagnon men of Cro Magnon were found in 1868 by an Englishman and Frenchman working together. Actually their fossils are neither the first Cromagnon men to have been found (others were found at nearby Aurignac in 1852 and are now beyond the reach of science, having been promptly reburied in a Christian graveyard), nor probably the oldest of the Cromagnon men, but they have given their name to a whole range of fossil remains that span a period of from, say, 30,000 to 10,000 years ago and have been dug up in an area extending from France and northern Spain to the Ukraine. Of course there is a considerable range of physical variation among this body of fossil men but they are sufficiently alike (and different from what had gone before) to take them together. They were strong men, with long arms and long legs and powerful musculature. When we look at their skulls and their easily reconstructed faces, we realise at once that we are looking at the modern sort of man – in fact, we are clearly looking at *European* faces from the start. Cromagnon skulls are lighter than any that went before, thinner-walled and with an altogether lighter-boned face. They are the first fossil men to display a really domed cranium and high forehead, and underneath their forehead and modern-looking brow they carry a face that looks pulled-in by comparison with Neanderthal and erectus faces, leaving the nose to stand out in relief from high cheek bones. Gerasimov has reconstructed both French and East European Cromagnons and his working methods were again no different from those he employed on modern missing persons' skulls for the Russian police. His reconstruction of the man from Combe Capelle in France (probably one of the oldest Cromagnons) shows his subject wearing the sort of bone adornment often found by archaeologists in association with Cromagnon interments.

The Cromagnons were accomplished hunters who manufactured and used a sophisticated range of stone and bone and antler tools and weapons. They built themselves hide tents and houses where they did not have caves to live in, they fashioned elaborate clothing against the cold of the final phases of the last Ice Age in which they lived and they decorated their clothes and themselves with beads and pendants and head-dresses. They made bone whistles and flutes and, in some of their deep caves, they danced in circles as the imprints of their heels in the clay cave-floors have revealed. And, of course, they initiated the long history of art: since then there have been no human cultures without some sort of artistic tradition. Unless the Neanderthals, as one anthropologist humorously suggested, brought say choral singing to a higher plane than has ever been reached since, we can be pretty sure that the Cromagnons were the first artists of any sort: certainly they were the first engravers, painters and sculptors. Most famously, they depicted on the walls of their caves (not their home shelters, but deep in their cult caves) the animals of the chase: the reindeer, horses, bison, mammoths they hunted to eat. Hunting magic seems clearly to be implicated in their art: there are instances of animals painted full of spears and even cases of modelled animal bodies that were actually speared by their Cromagnon creators. There is more to the oldest art than this: there was from the first a great emphasis on depicting the human, especially female, sexual organs – not exactly the Face of the Past, but in fact the subject matter of the world's earliest attempts at any sort of human representation in art. Piercing animals with spears and human sexual penetration seem sometimes to have been confused in the iconography of these first artists.

A separate line of artistic expression among the Cromagnons concerns predominantly bulky females in the form of the so-called 'Venus' human figurines. These small figures, a few inches high and carved in ivory or modelled in clay, come from sites again stretching across from France to Russia. Until a late phase of the Upper Palaeolithic period, these 'Venuses' constitute virtually the sole mode of human depiction in Cromagnon art. They are of an exaggeratedly sexual character, with an emphasis on breasts, belly and thighs to the near-total exclusion of other details. The 'Venuses' are the first indication that the Cromagnon artists, when it came to depicting human beings, saw themselves in an imaginative light quite removed from what we think of as the naturalism of their animal art. The truth is that their rendering of the animal world is not strictly naturalistic in any case: their bulls and bison are not photographically realistic but actually quite distorted in relative proportions of head and body and legs, for example. Their animal engravings, paintings and sculpture nevertheless satisfy our notions of a fundamental naturalism: their human depictions do not.

A review of the 'Venus' figurines shows that not all of them evince a positively obese femininity, although very many do. The 'Venus' of Ostrava-Petrkovice in Czechoslovakia resembles, in its broken state, one of the more presentable pieces of modern civic female nude sculpture – neither the breasts nor belly are exaggerated, but there are no head or feet. The 'Venus' of Willendorf in Austria, on the other hand, possesses breasts so large that her tiny arms can only lie across the top of them: under her stylised head of hair, she shows no face at all. The 'Venus' of Gagarino in the USSR and a 'Venus' from Dolni Vestonice in Czechoslovakia are very fat and the artist, as usual with these figurines, had no interest in their faces. The 'Venus' of Lespugue in France at least possesses a long neck. 'Venus' figures were also carved in some French caves – one of them, again a trifle overweight, holds up a bison horn like some modest cornucopia, but the artist has not wanted to introduce any detail at all into her face.

When the Cromagnon artists, after about 15,000 years ago in the final millennia of the Upper Palaeolithic, occasionally painted and engraved what we take to be human figures among the animals on their cave-walls, they resorted to a frankly cartoon-like caricature: so much so that the best that can be said of most of these representations is that they are 'anthropomorphic' – human-shaped rather than certainly human. Of course, the Cromagnon artists knew their own business best and clearly wanted to depict their human subjects in this way. Their caricature usually tends in an animal direction (with turned-up snouty faces like the representations from various caves in the French Vienne) or actually depicts human beings in animal guise (like the series of figures apparently wearing animal skins engraved at Les Combarelles in Les Eyzies or the deer-pelted and antlered figure or the 'sorcerer' painted at Les Trois Frères in the Pyrénées). It has been suggested that Cromagnon Man so valued and identified himself with the animal world that he not only wanted to 'animalise' his portrait of himself but could not help doing so. The animal world, whose individual members so instinctively conform to pattern and do not change their minds and ways like talking, thinking human beings, seems more eternal and more fundamental than even the conservative human world of the hunters.

The engravings and paintings of Cromagnon men include rather bird-headed human figures at places like Casares in Spain and Lascaux in France, and strangely bearded bison-faces at Le Portel and Les Trois Frères in the Pyrénees and Gabillou in the Dordogne. From the French cave-site of Marsoulas comes a rare frontal rather than profile human face, incoherently mis-shapen with an impossibly wide nose, the whole conveying once more an animal effect. More human but very stylised and simplified are the tiny carved ivory faces of a man from Dolni Vestonice in Czechoslovakia and a woman from Brassempouy in France.

The art of the Cromagnons vanished with the end of the rich hunting-life of its creators at the end of the Ice Age. In France, where once the fabulous cave-paintings of Lascaux had been created, the hunter-gatherers of post-glacial times were reduced to painting simple designs on pebbles. Some post-glacial communities in Denmark produced little engravings on bone of human figures occasionally similar in their gross stylisation to the Upper Palaeolithic art. In the Spanish Levant there is a post-glacial rock-art (with an unknown relationship, if any, to the Ice Age art) that shows in a very stylised way homely scenes of gathering honey and going out hunting. By this time, Cromagnon Man had at least 15,000 years behind him and Homo sapiens sapiens, whose oldest known representative he is, was a global species with members on all continents (except Antarctica), including Australia, which he had entered by perhaps 25,000 BC, and the Americas, where he was certainly established by 12,000 BC and perhaps earlier.

Obviously Homo sapiens sapiens did not arrive everywhere at once: he was almost certainly at home in Europe in the shape of the Cromagnons by 35,000 years ago and he may even have been established elsewhere before that (though the evidence is lacking). In the rest of the world there may have been survivals of more Neanderthaloid forms for considerable lengths of time; in Australia there is evidence to suggest that even erectus traits were still around at a very late date of only a few thousand years BC. All these considerations invite an enquiry into the origins of the present-day races of the world. Quite apart from racist prejudices, the subject is a difficult and obscure one – for the fossil evidence is largely lacking in several key areas and the situation of historical times is often puzzling in the extreme.

The Cromagnons are clearly Caucasoid in race: they are Europeans of a generalised ancestral sort. In the early years of the century, it was fashionable to see in the Cromagnons the ancestry of all the races of the world. There were thought, for example, to be Negroid and Australoid traits present in some of the Cromagnon burials, with the implication that descendants of these particular people had populated the parts of the world remote from Europe. This approach to racial origins continues fundamentally unchanged in the school of thought that has all the races descending from the first population of Homo sapiens sapiens, whether this population first evolved in Europe with the Cromagnons or – more likely on this view – evolved elsewhere and first appears in the fossil record as the Cromagnons at a time when Homo sapiens sapiens was already beginning his world-wide spread. Forty thousand years or so are perhaps long enough for the initially small groups of modern men as they trekked to far-away places to adapt some generalised H. sapiens sapiens appearance into local forms – and indeed some of the racially distinct traits of man do seem to be environmentally adapted, like the long, lanky and heat-dissipating bodies of the Negros of the Sudan and the short, compact heat-conserving bodies of the Eskimos.

Another view of the origins of races goes back to the anatomist Weidenreich who worked on Pekin Man; his work was greatly extended and revived by the American anthropologist Carleton S. Coon in the 1960s. On this view, the main racial divisions of man are older than

Homo sapiens and go back a million years to the first erectus populations. Weidenreich thought that Java Man already possessed traits which showed him to be a remote ancestor of the Australoid race which currently numbers among its members people living in South India and Papua New Guinea as well as the Australian aborigines. It certainly seems to be true that individuals of a markedly erectus appearance were living in Australia only approximately 10,000 years ago and the present-day Australian aborigines, though belonging to Homo sapiens sapiens, may preserve a few more 'primitive' features. Similarly, Coon proposes a Mongoloid racial lineage from the Pekin Man erectus specimens through some rather generalised Mongoloid individuals from later sites (contemporary perhaps with the generalised Caucasoids of the European Upper Palaeolithic) to the Mongoloid peoples of today. The Caucasoid line might be traced back from the Cromagnons through Swanscombe and Steinheim to Heidelberg Man. The Capoids of southern Africa, who include the Bushmen and the Hottentots, are only very partially traceable in the fossil record and the Congoids (or Negroids) of Africa are hardly traceable at all, having seemingly lived, before they spread east and south in Africa in quite recent times, in an area of West Africa where fossils were very unlikely to be left behind for us to find. The problem of Negro origins is very much further complicated by their apparent kinship with the Negroid and Negrito peoples of India, South East Asia and the Pacific Islands (there seem to be other strains in these places too): no historical or archaeological link can be discerned between these widely separated groups. Coon proposes that all the racial groups of Homo have come down since erectus times through the same stages of evolution with increasing brain-size, not always simultaneously all over the world but not in total genetic isolation from each other either, so that both by processes of convergent evolution and genetic interchange the racial groups have arrived at the same sort of H. sapiens sapiens status without losing their racial distinctiveness at the same time.

Whether it all happened within the last few tens of thousands of years or whether the process of racial formation has been a very long one, there are vast tracts of change and development here which we can barely glimpse at all. There have been thousands, millions of Faces of the Past into which we can never peer: if we could, we should perhaps be able to identify the first hints of the main racial divisions of the world today and we should perhaps be mildly surprised from time to time at unexpected mixtures of features and colourings. We can be sure that, since the establishment of Homo sapiens sapiens over most of this Earth, we would not have met with alien faces among any of our human ancestors. And similarly, for the rest of our story, we shall not expect to look into any very strange and exotic faces: they may have painted themselves or covered themselves with tattoos or worn amazing masks and head-dresses, but no physically outlandish faces of the sort of Australopithecus or Homo erectus or Neanderthal Man will appear again when we look into the Face of the Past.

Chapter Seven

From the First Farmers to the First Civilisations

10,000–2000 BC

With the world-wide arrival of Homo sapiens sapiens the physical evolution of mankind seems to have stopped. In fact, it is perhaps still going on but at a pace we cannot discern any more than individual Pekin men or Neanderthalers would have noticed it. Current theories of evolution suggest that all biological evolution has, in any case, proceeded not in smoothly gradual fashion but via plateaux of stability without much alteration, punctuated by swift bursts of change in response to enviromental opportunities. Human evolution is, moreover, subject to the effects of culture, which no other organisms possess. Technology, language, traditional ways of doing things all display their own courses of evolution and facilitate, when necessary, adaptive modifications to the environment far faster than any purely biological mechanism could offer. For example, human beings have developed hide clothing, diving suits and even protective wear for Outer Space to adapt themselves to environments that purely biological adaptation would take at least tens of thousands of years to achieve, if ever. In the light of technology, especially the high technology at whose threshold we now stand, some people have speculated that there may be no further need of any biological evolution: technological assistance, perhaps even some sort of union between man and the artificial intelligences he may one day create, will perhaps suffice to carry on the evolution of mankind. Or perhaps, after all, there is scope for some further increase in human brain-size in line with the whole course of human evolution to date. Even here, it is clear that the old biological mechanism – applicable to the rest of nature and to the early stages of human evolution – of random mutation and natural selection will not be operative, since, if it is to happen, the 'improvement' of the human brain will be purposely engineered by man himself.

Be all that as it may, there is no doubt that the emphasis of the evolution of mankind has wholly shifted since the appearance of H. sapiens sapiens from *physical* evolution to what Sir Julian Huxley called 'psychosocial evolution'. This is the progress of man's social arrangements, his technological accomplishments, his ideas and beliefs and his art. Everything about him, in short, that is not directly determined by his physical make-up; but which sustains, as it were, a life of its own outside the physical reality of individual men in the form of traditionally inherited but always-changing patterns of technology, economy, society, art and belief. Let us here simply contrast the physical evolution of man with the subsequently elaborated evolution of his culture.

Of course there was a slow and limited evolution of human culture during the several millions of years of human physical evolution. In the archaeological record, we can trace the development of the crude pebble tools of the earliest hominids into the first course hand-axes which were later refined and supplemented, or supplanted with flake tools, and eventually finely-reworked blade tools. With the appearance of Neanderthal Man we have seen that the very first hints of human culture beyond the technological can be noticed in the shape of deliberate burial with grave-goods and traces of cult practices. The Cromagnons brilliantly pioneered the development of art itself with their cave-paintings and carvings and, in doing so, left us in no doubt of their fully-human propensity for religious belief and ritual.

From the archaeological remains of the Cromagnon hunters, their cave-deposits and houses, their tools and weapons, the bones of the animals they killed for food, we can arrive at a picture of their economic and social arrangements – especially in conjunction with the judicious use of evidence from modern hunting societies like, say, the Aborigines of Australia. Such societies of hunter-gatherers do not necessarily work very hard (in bursts of hunting and collecting rather than with the sustained drudgery of agricultural peasants or factory workers). They are not stratified into classes of 'haves' and 'have-nots' who command and comply, though they usually support 'chiefs' and 'medicine-men' and influential elders. They are very democratic in their conduct and the rôles of the sexes are not necessarily as polarised as they have been in more 'developed' societies. The members of such hunter-gatherer societies work when it is called for and rest to think and dream when it is not and they treat each other, including their chiefs and sorcerers, on terms of equality. Lacking any human despots to oppress them, they imagine a spirit-world susceptible to their manipulation without any high gods to be flattered and appeased. Their liabilities, which are manifold, include their near-total dependence on the immediate performance of the world of which they form an integral part and their inability, other than in exceptional circumstances, to pile up a surplus of food against constant need. What they can on occasions achieve when the living is easy and abundant is shown by, for example, the recently flourishing Indian societies of the north west coast of America, where settled life with fancy huts and totem poles was accompanied by lavish feasting and a cult of gift-giving and conspicuous consumption: all supported by the rich fishing possibilities of their home-patch. It looks as though the Cromagnons, of France and northern Spain in particular, were similarly well-placed. But their way of life was terminated by the end of the Ice Age conditions which had brought huge migrating herds of game on to their doorsteps. Forests spread over their erstwhile hunting-grounds, the reindeer went north and those hunter-gatherer families who did not go north with them into altogether harder times stayed on in Europe to settle for a meaner life of hunting red-deer and fishing in the creeks and marshes of the post-glacial world. They produced very little art.

The end of the Ice Age, about 10,000 years ago, brought altered conditions of life all over the Earth. And these altered conditions seem, in one particular area, to have constituted the stimulus that set mankind upon the course of an environmental adaptation, the like of which the world had never seen before. The adaptation – farming, in a word – was wholly cultural in nature, a product of man's inventiveness. Man did not have to change his physical nature, nor indeed improve his mental capacities, to effect this adaptation: he simply used his powers of observation and his gift of invention. It seems certain that only Homo sapiens sapiens could have done it: for the very same environmental stimulus must have occurred at the end of the previous phase of the Ice Age, about 100,000 years earlier, but the late erectus/early sapiens representatives of humanity living at that time did not develop the farming life – probably because it simply could not occur to them!

The oldest known development of agriculture occurred in the Near East, in an area called the Fertile Crescent, arching from the Nile Valley through Palestine and Southern Anatolia to Mesopotamia, with an eastward extension to the Zagros Mountains. In the last millennia of the Ice Age, this area was more open and less fully forested than it was to become in post-glacial times. The present-day pattern of Mediterranean vegetation was restricted to pockets during the Ice Age and in the first millennia after the retreat of the glaciers in the north. It was only when post-glacial levels of winter rainfall were established that the modern pattern of Mediterranean vegetation spread (for a time more widely than today) over the region. That vegetation included the large-seeded grasses that are the wild ancestors of wheat and barley. In the seeds of these grasses, able to withstand summer drought, an attractive source of food was stored for the gatherers of post-glacial times. And at the same time, across North Africa down into Egypt and on through Palestine and Syria to the Zagros, there was a tradition of hunting the herd animals that were the wild prototypes of domesticated sheep and goats, cattle and pigs. None of this gathering of seeds and hunting of herds was actual farming practice, of course, but it prepared the way. Among the remains of some groups of hunter-gatherers in Palestine and neighbouring areas, called Natufians by archaeologists, excavation has turned up sickle flints, for gathering-in the seed-bearing grasses, and mortars carved in stone for pounding up the grain. Traces of Natufian huts and cemeteries indicate that the gathering of these grasses made possible longish periods of settled village life. The Natufians seem, moreover, to have deliberately seeded places where the stands of grasses did not naturally grow already: thus they initiated the cultivation of crops.

At Jericho, on the West Bank of the Jordan where Joshua was to achieve an important conquest in Biblical legend, the Natufians made a transition about 9000 years ago from the hunter-gatherer life to established village cultivation, with the beginning of the human breeding of food-plants into more convenient and higher-yield forms. Jericho is near to places where the wild ancestors of wheat still grow today but it was not a spot where that wheat grew by itself: it was, however, sited on a spring which still bubbles up near the mound. Progress was astonishingly rapid at Jericho once human settlement began. The spring was the source of the site's attraction and was evidently coveted, for Jericho boasts something more that had not been seen in the world before: stone architecture in the form of a tower 30 ft (9 m) high with an interior staircase, built to defend the village against attack from other human groups. It seems likely that those hostile outsiders were eventually successful, for there is a break in the archaeological record at Jericho when newcomers with different, rectangular houses arrive at about 6500 BC. Among their remains is the group of skulls found in one of the houses which may belong to revered ancestors of the community: their living features were restored on them in painted plaster and their eyes inlaid with shells. They are mostly clean-shaven faces but one has a moustache painted on: they offer a vivid image of what some of the first farmers looked like. There are round houses of mud-brick from the earliest stage of Jericho's long history and, though pottery was not yet being made, from the site come some of the distinctively polished stone axes that were once thought to be the very hall-mark of the Neolithic Period, the New Stone Age when the chipped flint axes of the Palaeolithic give way to polished forms. The Neolithic Period is now firmly associated with the beginnings of farming, and archaeologists speak of 'The Neolithic Revolution' to mark this time as a period of change every bit as important for the human race as the Industrial Revolution of nineteenth-century Europe.

For the invention of farming brought an alteration of human existence more far-reaching than any other human invention until the present day. When the domestication of animals was added to the cultivation of plants, the Neolithic societies of the ancient Near East were able for the first time to regulate their food supply all the year round and to increase production with more manpower, with technological aids like the stone hoe and with improvements in their crops and stock by selective breeding.

It seems that for a long time, while the farming way of life was becoming established and still called upon all the members of society to labour together in the fields, something of the old egalitarianism and 'democracy' of the hunter-gatherers was maintained. But as the first few millennia of the new way went by, and more plants and animals were added to the farming repertoire with an increase in yield and the provision of real surpluses to be stored and administered, divisions of wealth and power became more marked. Whereas the hunters had worked hard together from time to time and the first farmers had worked very hard together most of the time, now there were landless peasants working drudge-hard all the time and land-owners (in effect) directing them, administering their production and living off their labour. By this time, about 6000 BC, farming had spread to coastal Mediterranean sites in the Levant that underlie such later historical cities as Byblos (part of modern Beirut) and Ugarit; some east Mediterranean islands had been colonised for farming (like Cyprus) and parts of the Greek mainland and southern Anatolia (in Turkey). There were seafaring and trade in the fine volcanic glass called obsidian that made such good tools. Mud-brick architecture could now encompass multi-storey buildings and cattle had been added to the range of domesticates. There is a site in Turkey called Çatal Hüyük (Hüyük means the same as Tell – both indicate those mound sites created by the rebuilding of ancient cities on the same spot over thousands of years) where an elaborate complex of buildings includes 'shrines' decorated with real and remodelled bulls' heads which clearly played a part in some cult of the bull. From Çatal Hüyük too comes an extraordinary wall-painting that shows what is believed to be the face of a dead man, stylised in a way resembling children's painting: one of the strangest Faces of the Past in ancient art. The iconography of Çatal Hüyük includes such other sinister elements as vultures with human legs attacking headless men and there is, among many depictions of a 'mother-goddess', a very fat figure apparently giving birth. A site in northern Mesopotamia reveals a Neolithic village of the sixth millenium BC with store-rooms that indicate the provision of food surpluses and demarcated 'industrial' zones that point to divisions of society with craft-specialisation. The craft in question was pottery – the firing of pots had now been added to the technological resources of the Neolithic. The first evidence for the smelting of copper out of its rock appears at the same time. In Mesopotamia, villages became bigger and more complex in the sixth millenium, with simple irrigation of farming land and the greater preparation of the fields with hoes. These villages possessed defensive walls and ditches. There was trade in luxury goods like copper and turquoise and the presence of seals and seal-impressions hint at ownership and authority: some people were getting richer and more powerful than the rest of their community.

Further south towards the Persian Gulf, and a little later on, the village communities that pre-date the appearance of the world's first civilisation – Sumer – were established on the basis of irrigation fed by the twin rivers Tigris and Euphrates. Irrigation made agriculture possible where it had not been possible before and promoted higher productivity: at the same time it called for a greater degree of co-operative labour and a more elaborate form of social organisation, with leaders and led. Irrigation-farming in Lower Mesopotamia was, additionally, a dicier business than earlier forms of agriculture: life was burdened with the dangers of flash-floods and dust-storms. It seems plausible to conclude that the increasing division of society into labouring and ruling classes, together with the increasing uncertainty of life in the face of storm and flood, turned men's minds towards a form of religious outlook more recognisably related to the great historical religions than the beliefs of the hunter-gatherers and the first egalitarian farmers. The Cromagnons' deep caves were their shrines, and their religion plainly involved fertility, whatever else it may have concerned in the way of hunting magic: the figures of no humanly-attributed gods appear in the cave-art and the fertility figures of the 'Venuses' do not merit the title 'goddess'. The first farmers fashioned fertility figures, too, often called 'mother goddesses', but even at Çatal Hüyük there is no trace of any almighty anthropomorphic god or clan of gods that we can discern.

In the Mesopotamian towns that underlie the great cities of the Sumerian civilisation, we find the first verions of temples that were to go on being rebuilt and extended on the same sites for thousands of years: after 4000 BC the towns were large and the temples monumental,

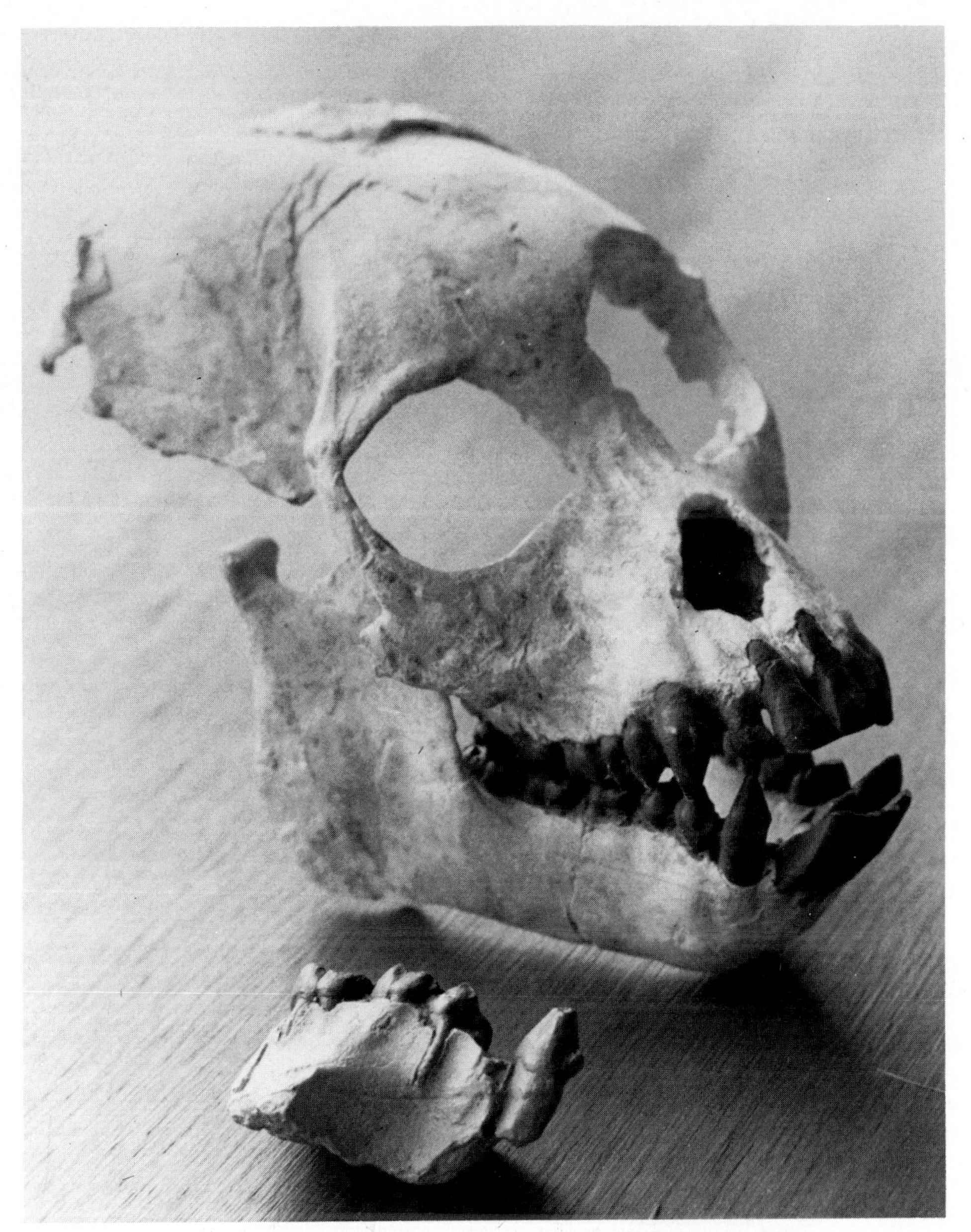

1 Ramapithecus: a jaw fragment and the skull of a closely related form from Africa, called 'Proconsul'

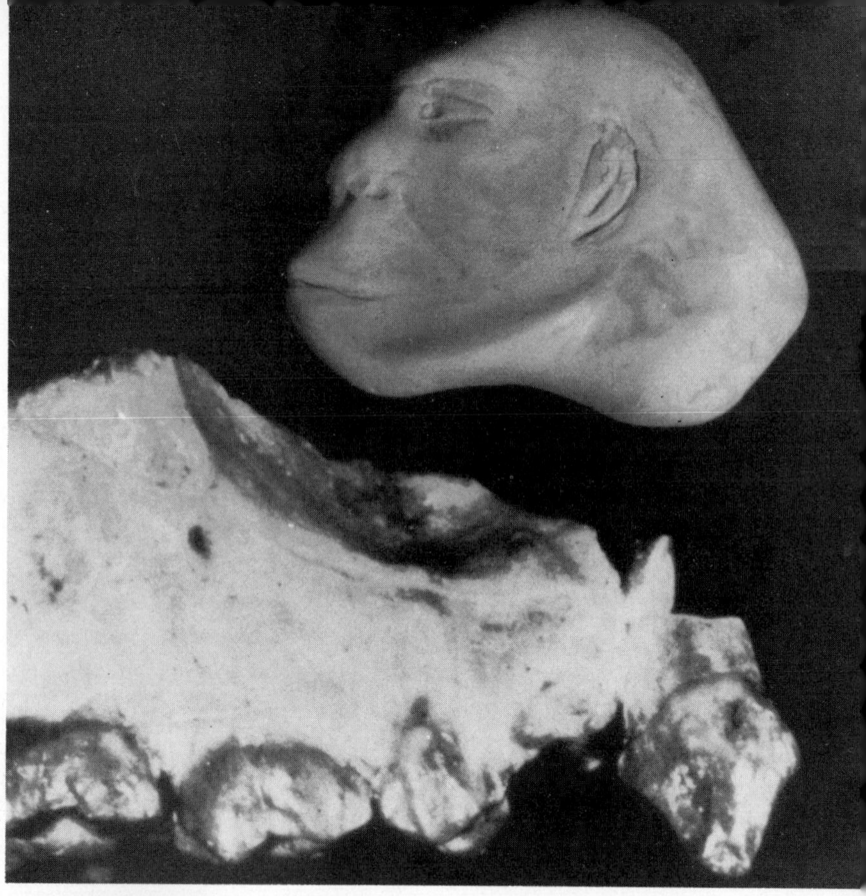

2 Ramapithecus: jaw fragment and speculative reconstruction

3 Haeckel's 'Pithecanthropus alalus' from nineteenth-century engraving

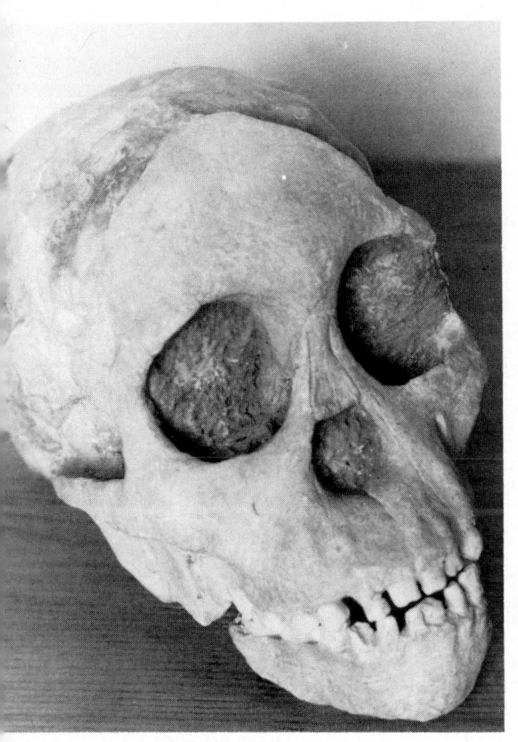

4 *The 'Taung Baby' skull*

5 *The 'Taung Baby's' face as reconstructed by Russian anatomist Gerasimov*

6 *A skull of Australopithecus robustus*

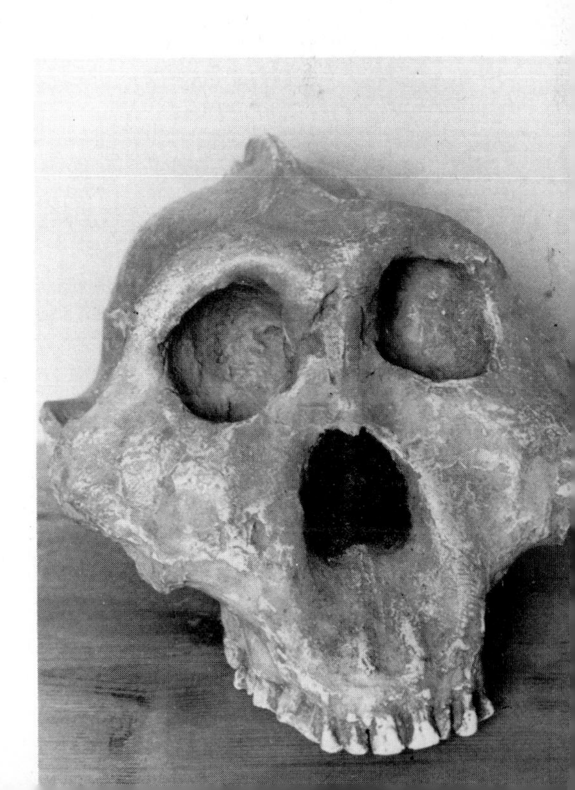

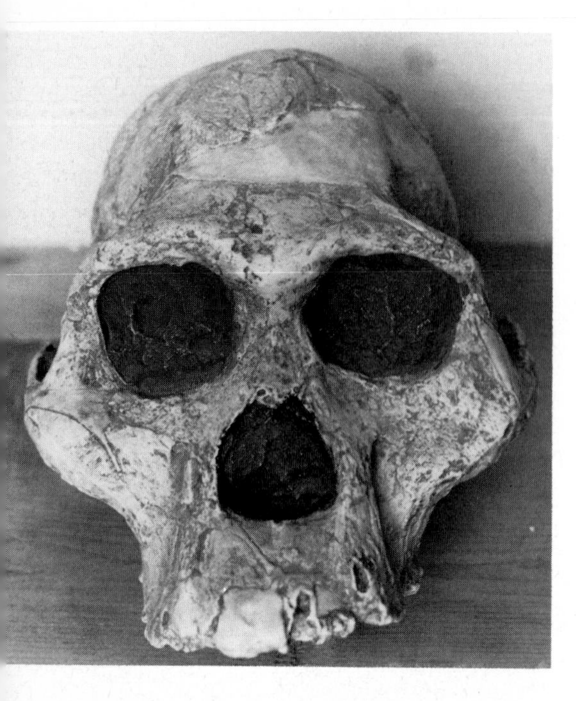

7 A skull of Australopithecus africanus

10 Opposite: Australopithecus hunters by Maurice Wilson

8 Australopithecus as reconstructed by British Museum artist Maurice Wilson

9 The face of Australopithecus africanus modelled by Maurice Wilson, without speculation as to hair and skin colour

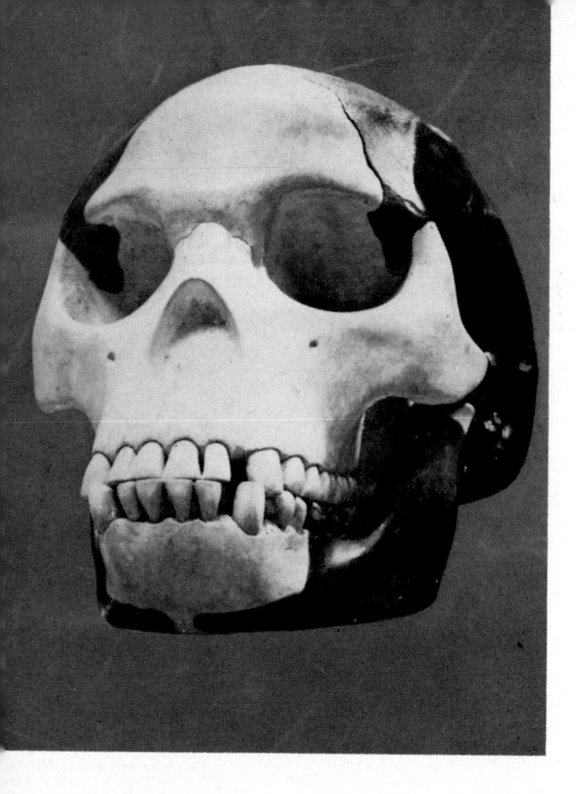

11 A reconstruction of the impossible skull of Piltdown Man made before the disclosure of the hoax

12 The noble brain-case and primitive jaw of Piltdown Man in doomed reconstruction

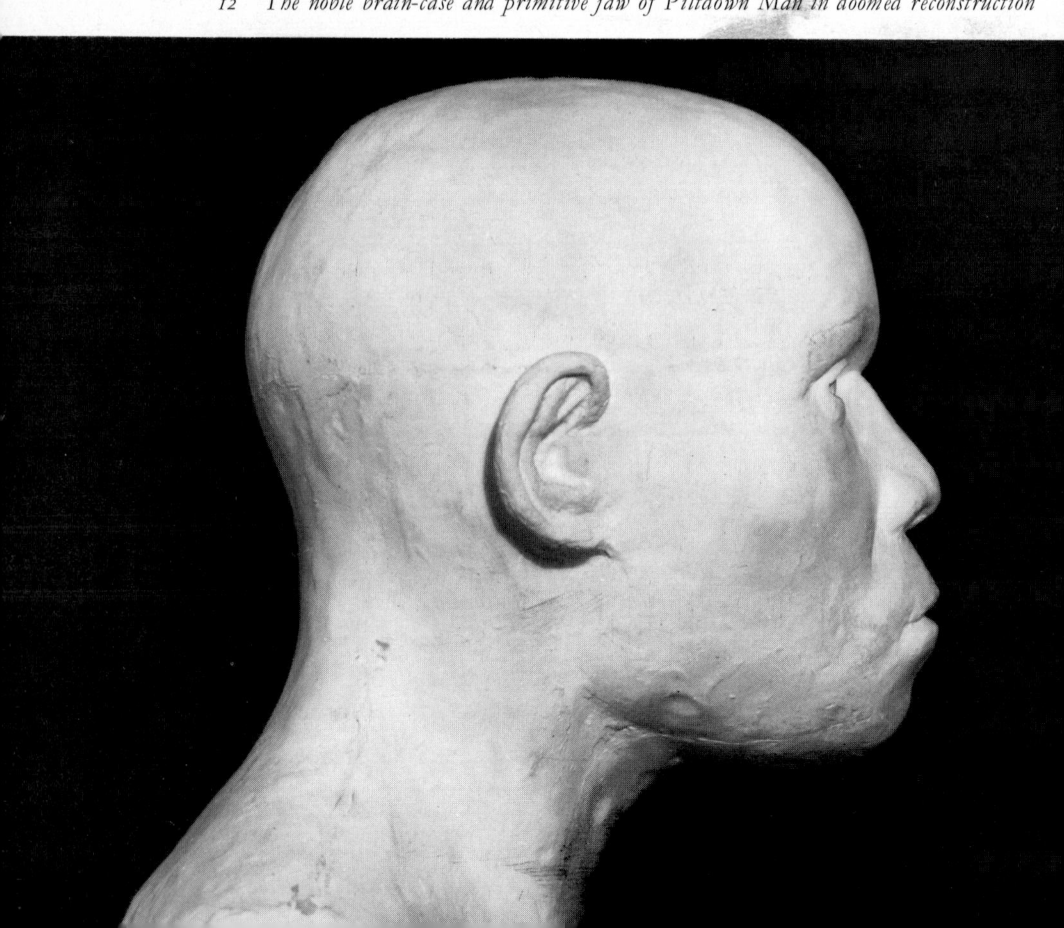

13 A skull fragment of Java Man (left) and skull of Pekin Man

14 The face of Java Man as reconstructed by Maurice Wilson

15 Pekin Man by Maurice Wilson

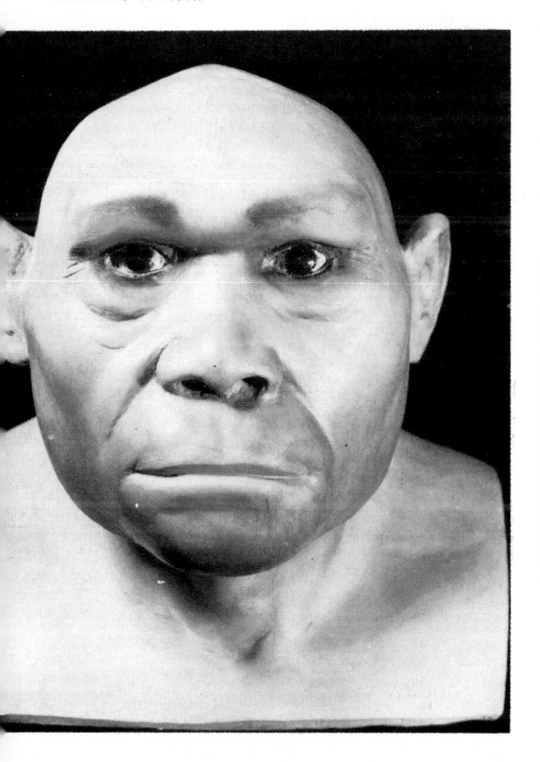

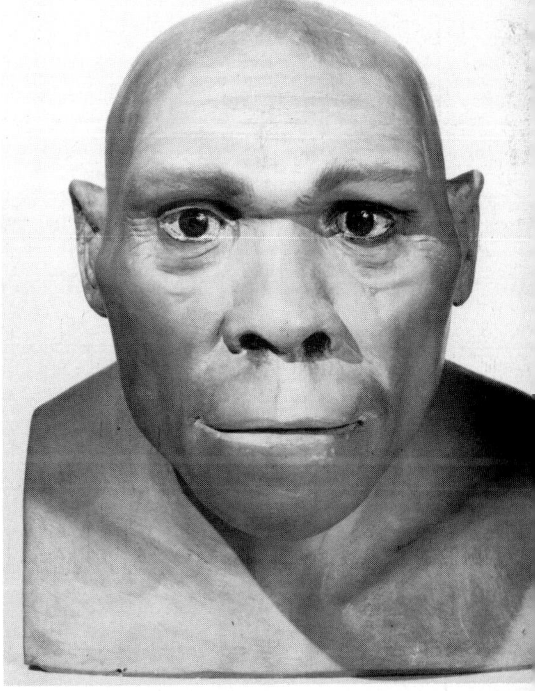

16 Pekin Man sculpted, with hair, by Gerasimov

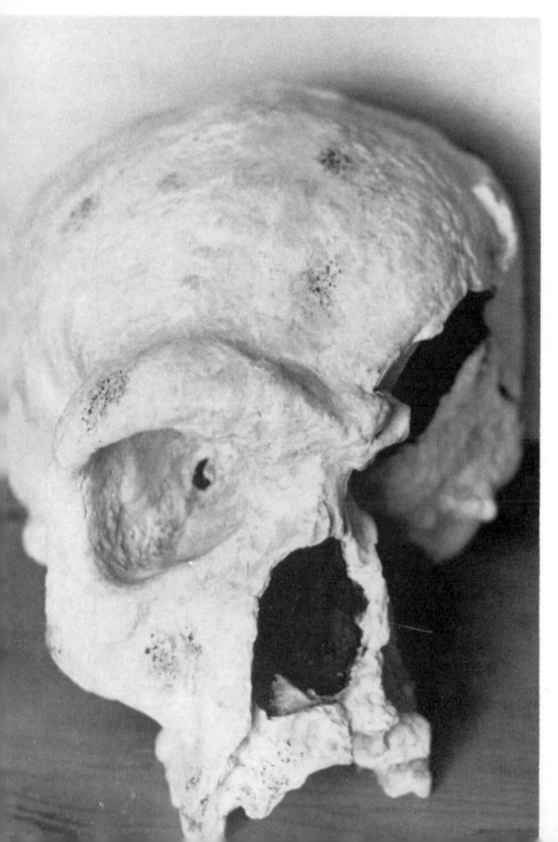

17 The skull of the Steinheim woman

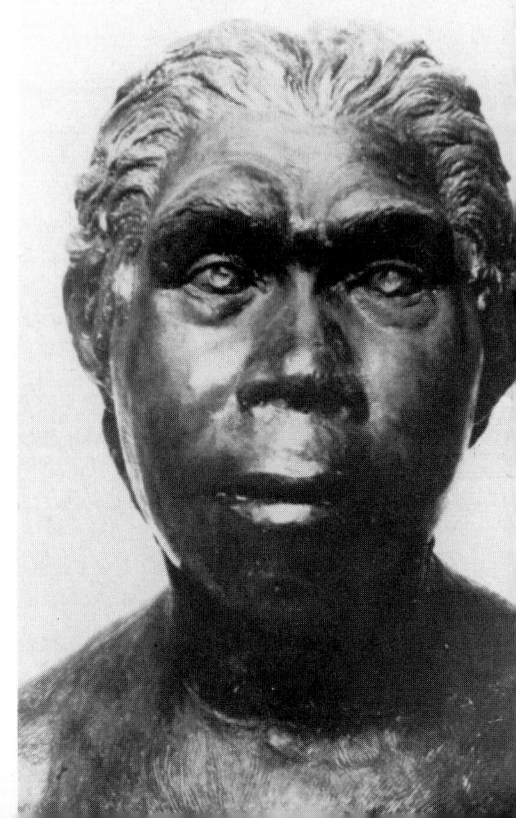

18 The Steinheim woman, by Gerasimov

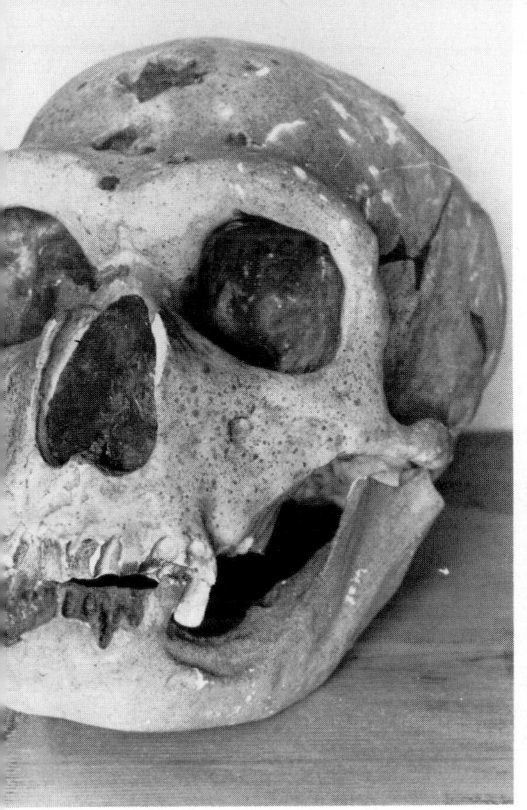

19 A classic Neanderthal skull

20 Neanderthal Man as reconstructed by Gerasimov

21 A Neanderthal child reconstructed by Gerasimov

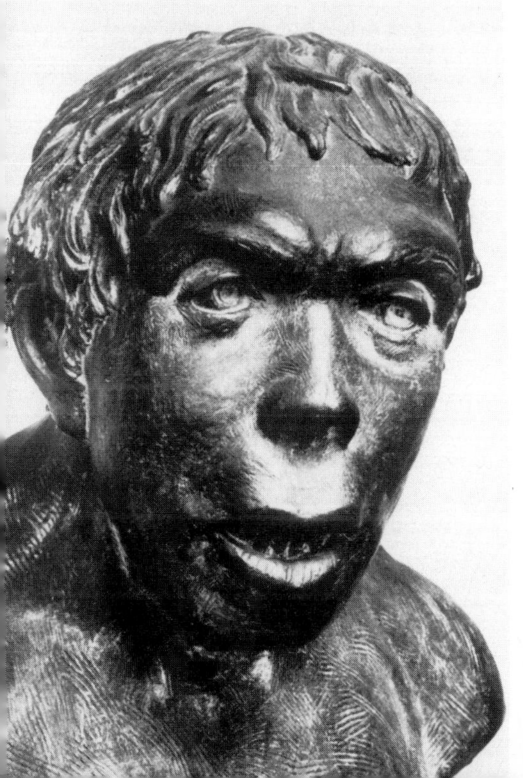

22 An adult Neanderthal male by Gerasimov

23 The 'Body-in-the-Ice' fake

24 A skull of a Cromagnon man

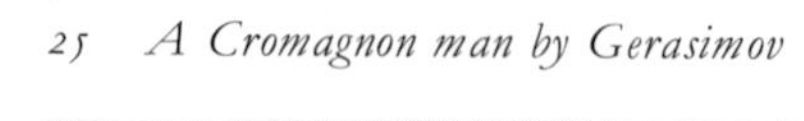

25 A Cromagnon man by Gerasimov

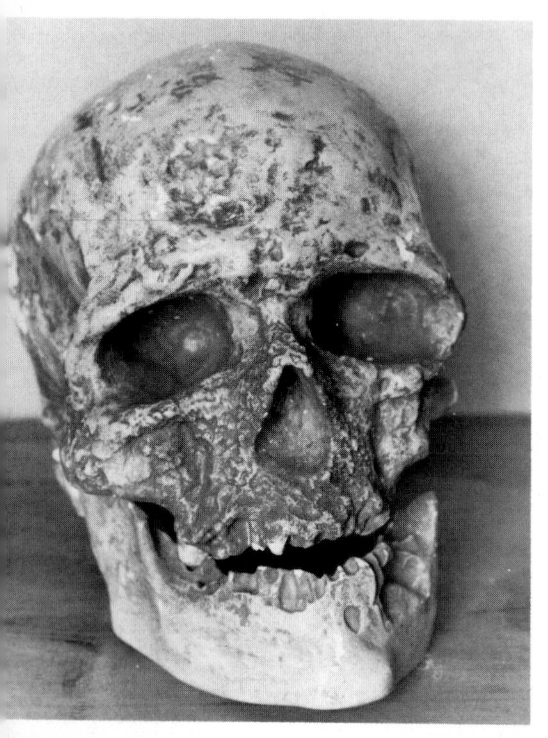

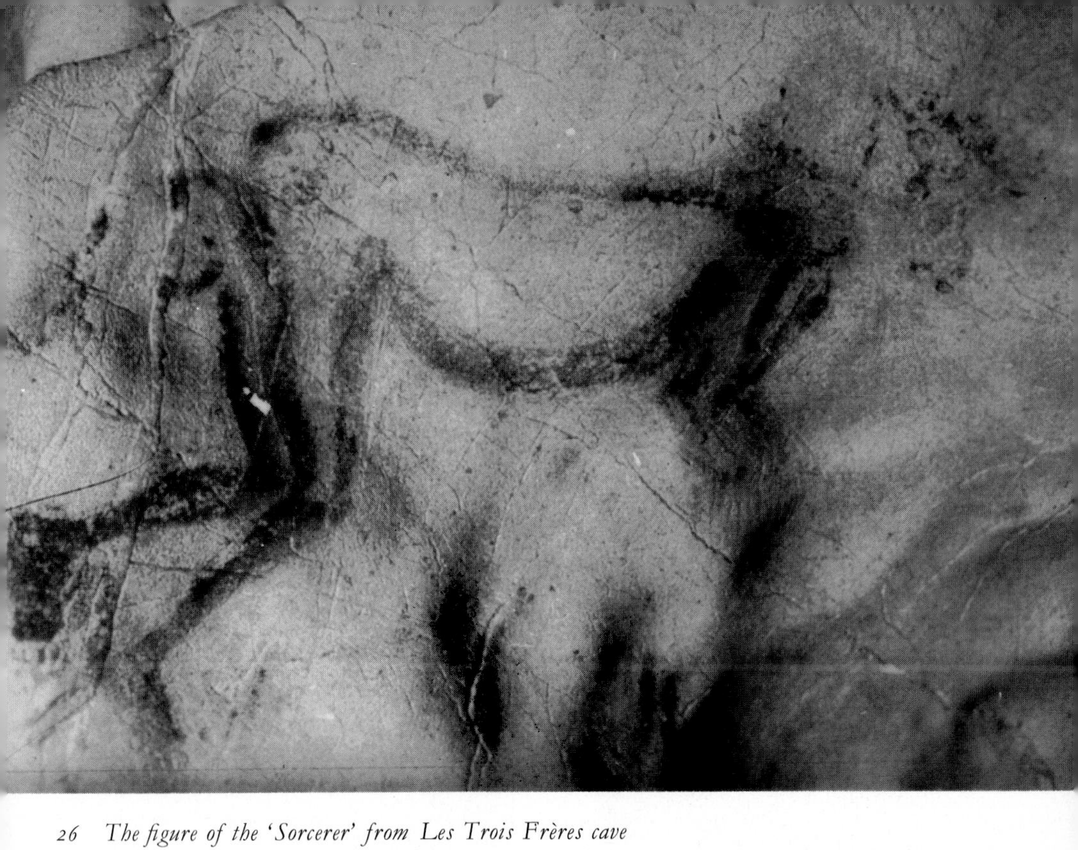

26 The figure of the 'Sorcerer' from Les Trois Frères cave

27 A 'Venus' figurine from the last Ice Age

28 The grotesque frontal engraving of a human face from Marsoulas

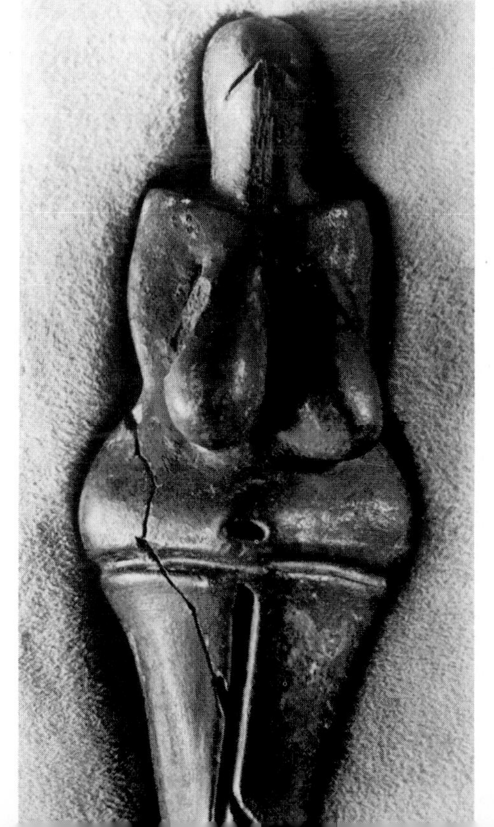

29 The tiny carving of a female face from Brassempouy

30 A fine 'plastered skull' from Jericho

31 The 'Dead Man's Face' from Catal Hüyük

32 Opposite: The so-called 'head of Sargon of Akkad'

33 The body of a predynastic Egyptian in the British Museum – called 'Ginger'

34 The mummy in the tomb of Nefer

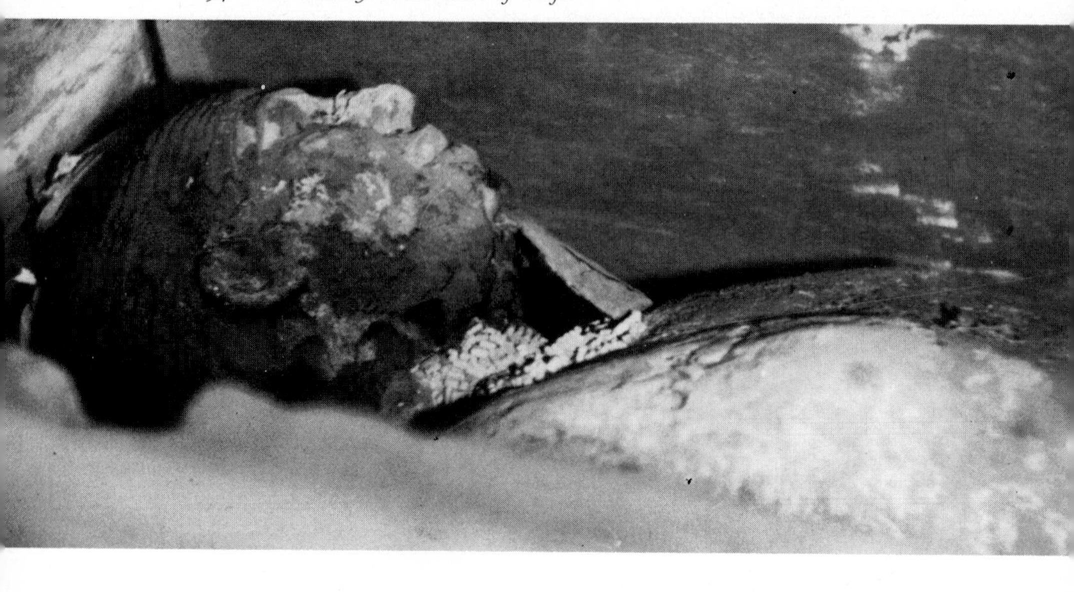

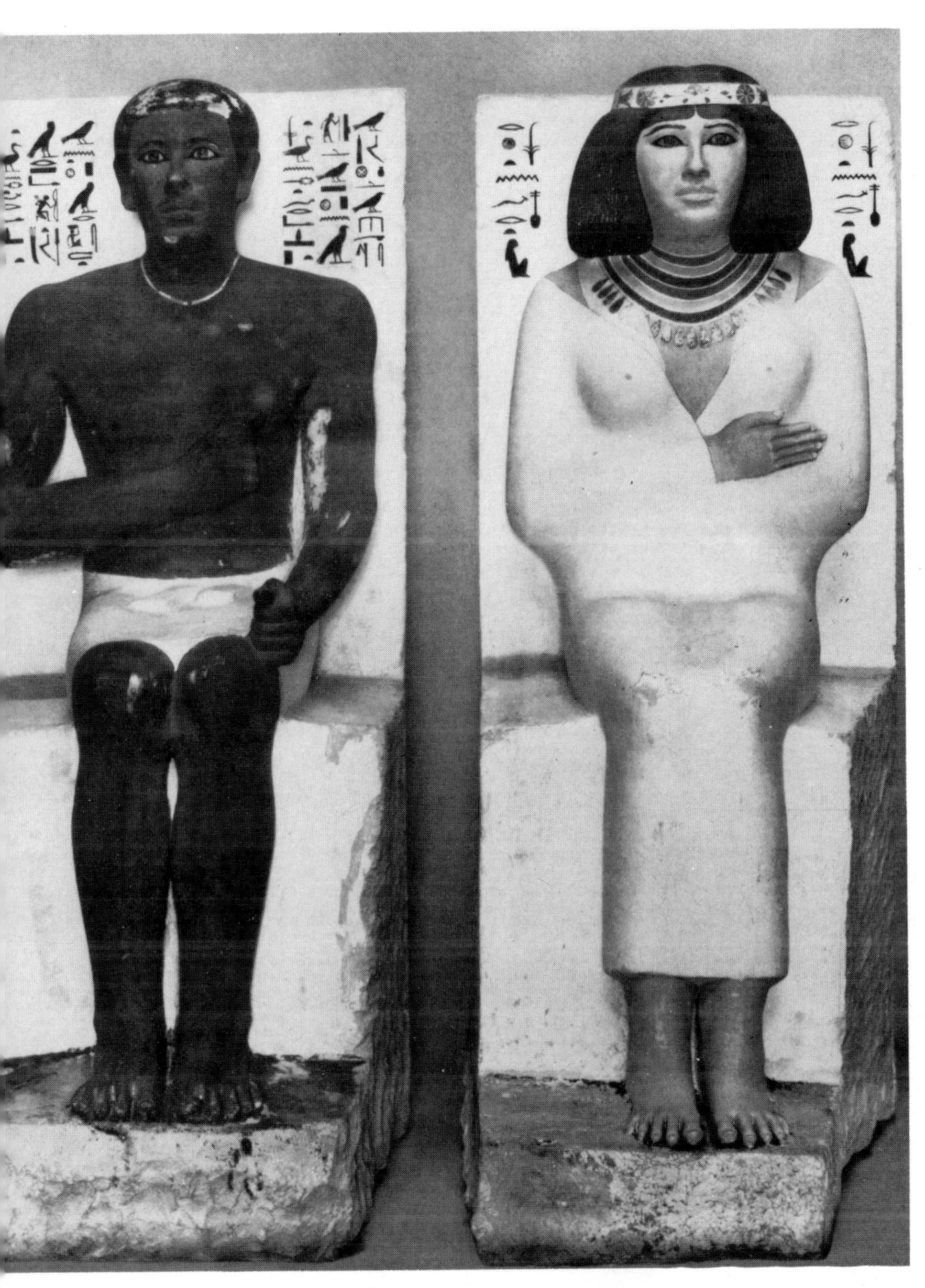

35 *Statues of Rahotep and his wife*

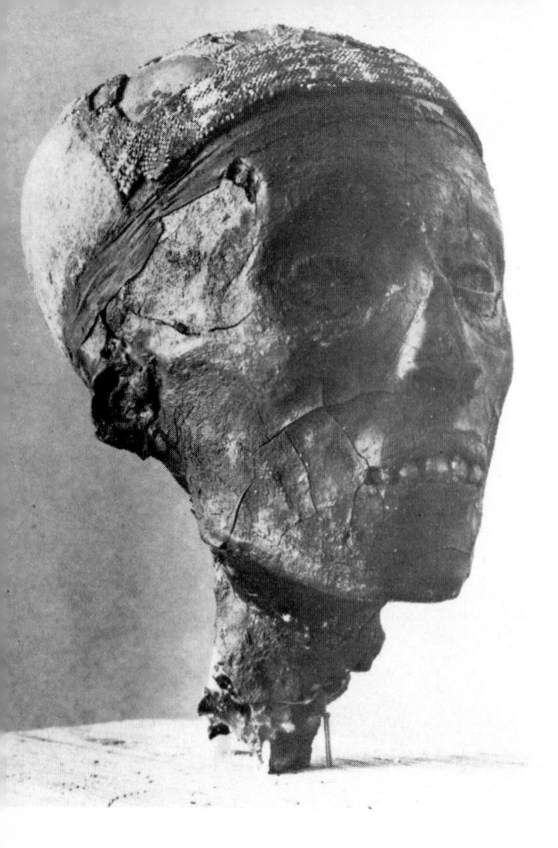

36 The head of Tutankhamun

37 The wounded head of Sekenenre

38 The mummy sometimes ascribed to Queen Hatshepsut

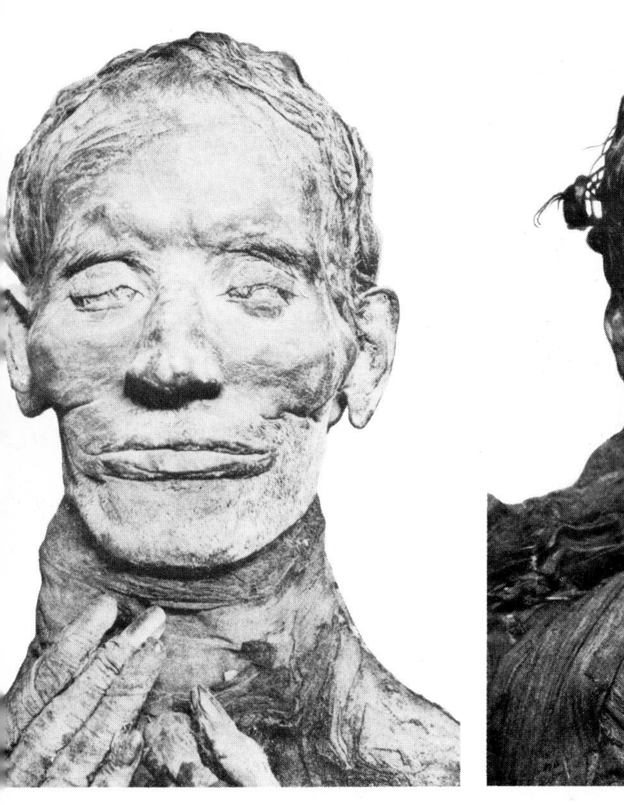

39 The face of Yuya

40 The face of Thuya

41 The official face of Akhenaten

43 Manchester mummy No. 1770 at an early stage of unwrapping

42 Left: Ramesses II in formal statue and in death

44 The fully reconstructed head of Manchester mummy No. 1770

45 A fine portrait-board on a mummy of the Graeco-Roman period

46 The mask of Agamemnon

47 The 'Archaic Smile'

48 Hellenistic art – a drunken old woman

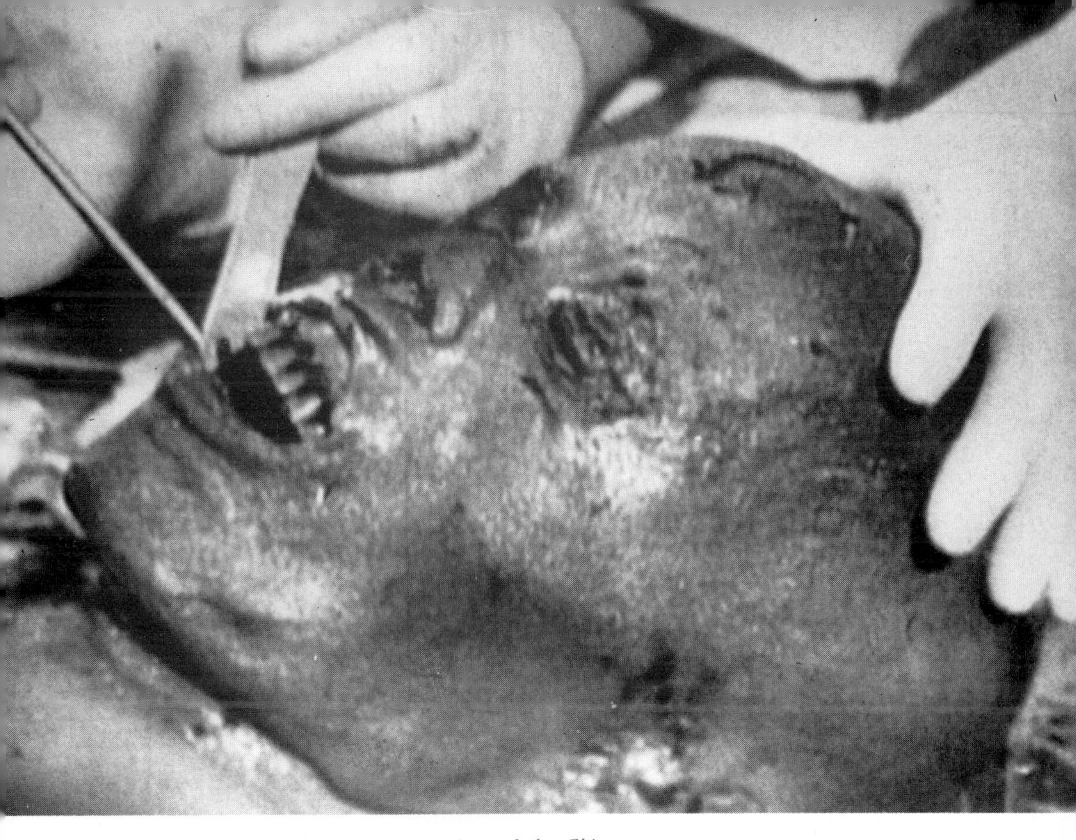

49 The face of Sui, during the investigations of the Chinese team

50 An Olmec head from Mexico

51 A Roman portrait sculpture of Julius Caesar

52 A victim of the volcanic eruption that destroyed Pompeii and Herculaneum

53 A fourth-century glass panel 'portrait' of Homer from Corinth

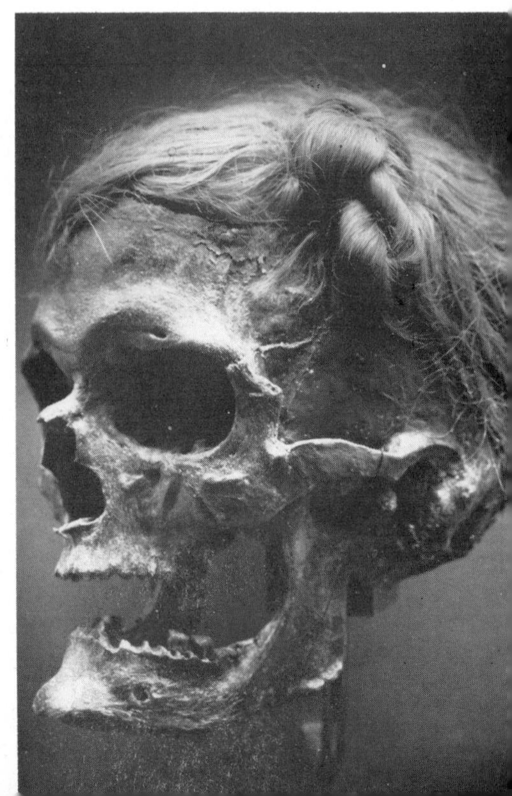

54 The Swabian side-knot on a bog-body skull

55 *The blindfolded face of the Windeby girl*

56 *The face of the Tollund Man*

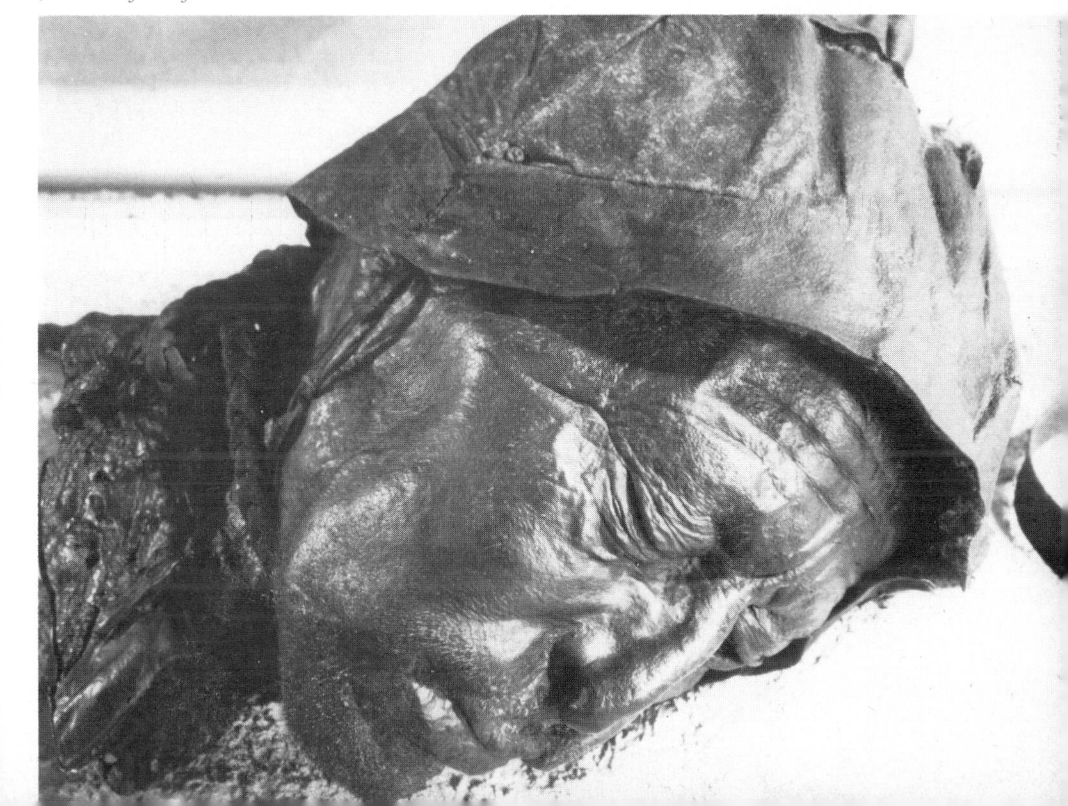

57 *The face on the 'Turin Shroud'*

dedicated to the service and flattery of gods patterned on the human rulers of the Mesopotamian city-states. When, soon after 3500 BC, the first written records of Sumer appear in the form of pictographic records of land-sales and commercial transactions of commodities, there are at first no 'words' for 'king' or 'palace' but there are signs hinting at the authority rôles of later times. And there is representational art depicting 'priests' or 'chiefs'. The temples were the indivisibly ceremonial and administrative centres of Sumer, the repositories of the gods' wealth (the food and trade surplus created by the labour of the gods' servants – the real human individuals of these first civilised communities) and the seat of the priestly power that regulated these communities. In fact, there was always a strain of church-state rivalry in Sumer and its successor-states of Mesopotamia, for the kings of this region were never decked in the divine authority of the Egyptian pharaohs. The earliest Sumerian representational art shows its human subjects from the first in a stylised manner that was to persist throughout ancient Mesopotamian history. Even in its highest achievements, like the white marble head from Uruk dated to about 3100 BC, the features of the lower part of the face are sensitively rendered, but the eyes and eyebrows are stylised in a direction away from realism; some of the other sculpture of the time displays that frankly 'ethnographic' look of ancient Mesopotamian art that contrasts with the naturalism of Egyptian art almost from the first.

It seems likely that the initial impetus towards representational art and writing in ancient Egypt was supplied by trade contacts with Sumer and the Sumerian sphere of influence. Some of the earliest pieces of Egyptian art, at about 3100 BC, have a markedly Sumerian look about them and the first Egyptian pictographic writing, perpetrated at about the same time and thus a few centuries after the first pictograms of Sumer, was probably inspired by the Sumerian example. But from the first, Egyptian writing used very different symbols from the Sumerian ones and rapidly developed its own character. Both Egyptian and Sumerian writing had lost their purely pictographic character after 3000 BC, no longer just meaning something by showing a picture of it but expressing the sound-content of the two mature languages. The Sumerian script went on to be adapted to the writing of another, Semitic, tongue when the Semite Akkadians came to power in Mesopotamia. And just as the Egyptian writing exhibited a character all of its own from the very beginning, so Egyptian art very soon shed its Mesopotamian residue and took on that distinctively ancient Egyptian look that combines strong stylisation with a definite humanely naturalistic quality. Though the kings of Egypt, from the First Dynasty of about 3100 BC when the two lands of Upper and Lower Egypt were united, were from the first regarded as divine figures who headed every aspect of ancient Egyptian life, they were almost from the first depicted as recognisable human beings and not as stylised religious symbols. From the first, too, they were buried in a distinctively Egyptian way – as mummies. It is possible that mummification was carried out in other parts of the ancient world, especially in places like Byblos, that were in close contact with ancient Egypt, but nowhere else had the practice of mummification both a natural origin and an ideal situation for preservation. People buried in the hot, drying sand of Egypt became naturally mummified in predynastic times and, when the Egyptians of historical dynastic times, took to burying their dead in deliberately mummified form, they committed their mummies to a soil that could preserve them like no other in the world, except in the odd exceptional cases that have given us the Chinese bodies and the bog-bodies of north west Europe. The Egyptian mummies afford us the chance to gaze into scores of the real Faces of the Past which, in other ancient cultures, we can only discern in representational art. What the Egyptian mummies lack in the sort of vivid living freshness with which the modern American morticians are able even to improve on life, the sculpture and paintings of the other ancient cultures (and of the ancient Egyptians, too, of course) make up for in perceptiveness and imagination – at the risk of idealisation and distortion away from the real appearance of their subjects.

While Mesopotamia and Egypt were making their ways towards literate civilisation by 3000 BC, the preceding millennia had seen the spread of the farming way of life beyond the Fertile Crescent into the whole Mediterranean region, into first central and then all Europe including Scandinavia, into India and the Far East. There may have been independent inventions of the farming economy in some of these places: it seems certain that agriculture was independently invented in the Americas, where the beginnings of cultivation had appeared in Mexico by 6000 BC. Sub-Saharan Africa's first agriculturalists probably date to after 2500 BC. Whereas it was once taken for granted that all human progress in every part of the world was inspired by the example of the pioneering cultures of the ancient Near East, archaeologists nowadays

credit human communities in all times and places with inventive powers capable of developing new ways of life for themselves without the necessary intervention of superior beings from outside. It is one of the revealing characteristics of the 'ancient-gods-and-spacemen' school of pseudo-archaeology (along with its staggering ignorance of the actual findings of real archaeology and idle indifference to its ignorance) that it has taken the nineteenth-century notion of cultural diffusionism to the ultimate absurdity, claiming that all human achievement – even the evolution of man himself – is owed to superior beings from Outer Space!
As the farming way of life was taken up in more and more parts of the world, its effects were in line with those manifested among the first farmers of the ancient Near East. A phase of Neolithic 'self-sufficiency' with hard work and long hours for all, was generally superseded by the emergence of class stratifications, with a more formalised drudgery of the peasant class and privileged leadership at the top.

After 3000 BC, by which time the civilisations of Sumer and Egypt were well established, the process of civilisation was accelerated and widely dispersed. In Mesopotamia, the city-states and dynasties of early historical times are well attested, and we can – through the medium of sculpture – look into the stylised and idealised face of a famed conqueror and ruler, the Akkadian King Sargon of about 2350 BC (though this piece may have been sculpted at a later date). At the turn of the millennium (about 2000 BC), the long-powerful city-state of Ur was eclipsed in times of ancient world-wide disturbance and unrest. Similarly tumultuous times brought chaos to ancient Egypt at the close of hundreds of years of continuity of the Egyptian Old Kingdom, when the famous pyramids were built and art reached a peak of expression alongside the perfection of techniques of mummification. During the same millennium after 3000 BC, the region of the east Mediterranean saw the establishment of towns and city-states in Palestine and Syria which traded with the islands and mainland cities of the Aegean: troubled times afflicted this region, too, around 2000 BC, when Indo-European language-speakers arrived in Anatolia and set off a train of disturbances.

The citizens of the east Mediterranean mainlands and islands have left us a representational portrait of themselves in the cult figures of gods and goddesses they produced for their shrines – heavily influenced by the Mesopotamian and Egyptian models with which they were familiar in the course of trade. In the west Mediterranean area there were villages trading with the east in things like copper, and here again a time of upsets had arrived by 2000 BC. In Europe, the spread of farming and village-life continued during the same third millennium BC, reaching even Scandinavia before 2000 BC, while on the Atlantic coasts of Europe from Spain to Denmark, the farming communities of these centuries were constructing the huge stone monuments known as megaliths that were frequently the burial places of the neolithic settlements as a whole or, as times went by, of their chiefs and priests. In the British Isles the first examples of more than a thousand stone-circles, of which Stonehenge is the most famous and elaborate, were started at this time.

In north west India, the Indus civilisation with well-planned cities and trading ports, was established after 2500 BC, to be brought down in troubled times at about 1800 BC when the Indo-European speakers were embarked on the conquest of the Indian sub-continent. The Indus civilisation has left to us some examples of an art distinctively its own, as witnessed by the lithe figure of a dancing girl from Mohenjo-daro and the stylised portrait of a bearded priest from Harappa. In the Far East, during the third millennium BC, the farming way of life spread across northern China while the hunter-gatherer style persisted in the south. In the Americas, hunters and gatherers predominated too, though the end of the millennium saw the establishment of settled farming life in Mexico and Peru.

It seems certain that most, if not all, of the human communities of the times of the first civilisations must have engaged in the creation of artistic efforts of some kind, alongside the production of their subsistence. Almost no human community has ever been charged with the failure to make some sort of art – probably not even the demoralised Tierra del Fuegans of recent historical times were wholly immune to artistic endeavour. Of course, there are many possible expressions of art that earn no chance of survival in the archaeological record: song and dance, for example (give or take a few foot impressions in a deep cave or on the land-surface beneath some heaped-up earthen barrow). But, in the right circumstances of preservation, wooden and more certainly ceramic and stone-carved works of art can survive, and from the well-established and long-lived ancient civilisations, like those of Egypt and Mesopotamia, such pieces have survived in abundance. Where such works of art more or less representationally depict the forms and faces of the human beings of their times, they offer us the opportunity to do what this book is all about: to look into the Face of the Past. But there is one source in the ancient world of the first civilisations that offers us a more direct means of insight and opens up a whole world of interest in itself: that source is provided by the mummies of ancient Egypt.

Chapter Eight

The Face of Ancient Egypt

3000–1300 BC

We can look into the faces of a great number of dead ancient Egyptians, thanks to the tradition of mummification which was nowhere more widely practised nor brought to a higher standard than it was in ancient Egypt. Other peoples at other times have created human mummies – for example, the Incas of fifteenth-century Peru and the prehistoric inhabitants of the Canary Islands – but the Egyptians systematised the practice and researched its possibilities to the limits of their resources.

In common with most other human communities, the ancient Egyptians from the very earliest times – well before the unification of their territory into a single state at about 3100 BC – evidently entertained the concept of an afterlife beyond death and burial. The graves of some of the inhabitants of the Nile Valley, who lived in the centuries of 'predynastic' cultures before the unification of ancient Egypt, have been discovered to contain, along with bodies of the deceased, grave-goods in the form of pots, beads, flints and so on which were almost certainly meant to serve the needs of the dead in the continuance of their lives after death. The graves of these predynastic people were usually shallow pits in the sand of the desert, just beyond the cultivated green fields of the Nile Valley which were presumably too precious to be polluted with interments. The bodies were flexed, rather then laid out, with the knees drawn up under the chin and the hands by the face. Buried thus in close contact with the hot, dry sands of Egypt, the bodies were rapidly desiccated, the fluids and products of decomposition drawn off and evaporated before severe corruption could set in. The ancient Egyptians of predynastic times must have been well acquainted with the naturally preservative action of the sand into which the graves of their dead were cut. A body accidentally turned up by animals or the extension of irrigation – dried out and shrunk but without loss of skin, nails, hair – must have impressed them as a more desirable 'survival' of death and burial than a disarticulated skeleton. Our museums contain examples today of predynastic burials of more than 5000 years ago: the British Museum has a particularly fine example popularly known as 'Ginger' on account of the coloration of his skin and hair as the result of the desiccation process.

The ancient Egyptians came to regard preservation of this sort as a religious necessity. Their religious outlook in any case emphasised the essential changelessness of things and set great store by eternity: they lived in natural circumstances which encouraged this attitude more than the environments of most peoples do, for sunlit Egypt exhibits a remarkably consistent geography for much of the length of the Nile river and even its changing seasons are exceptionally regular and dependable. It has been suggested that the ancient Egyptians' curious emphasis upon animal gods and animal iconography, upon which everyone since the classical travellers of Greece and Rome has commented with surprise, was owed to their perception that the vivid world of the animals presents a more unchanging and 'eternal' image than the world of men does. At any rate, from the first, the ancient Egyptians seem to have regarded the survival of the soul after death as being dependent on the survival of the body in its grave or tomb.

Ironically, it was the ambition of the ancient Egyptians in early dynastic times to furnish the bodies of their dead with costlier and more elaborate grave-goods in bigger and better graves that threatened to rob them of the high degree of preservation they had previously enjoyed in their shallow sandy graves. The sand-pit burials of predynastic Egypt were sometimes furnished with coverings for the corpse of goatskin or matting. Subsequently, the pits were roofed over with pieces of wood and matting, or at least a basket was placed over the head of the deceased. Later on came basketwork trays and covers which were the precursors of wooden coffins: the first coffins were boxes to contain contracted and not full-length burials. A stage was reached when the bodies of the dead were being buried in wooden coffins in rectangular wood-lined graves which approximated to the idea of 'houses' for the dead and, like the homes of the living, could house and protect the goods of the occupier against robbery – for the dead were now being equipped with richer grave-goods to serve them in the next world.

When the richer sections of late predynastic and early dynastic society in ancient Egypt started to elaborate the graves of their dead away from simple pits in the hot dry sand of the desert margins into structured tombs with wooden and stone compartments and chambers, the bodies of the deceased were removed from immediate contact with the very substance which had promoted their preservation previously – the sand which was able to draw off moisture and evaporate it, thereby slowing down and stopping the corruption. The ancient Egyptians of the few centuries before and around 3000 BC must have noticed the deterioration in their dead which the new burial practices had brought about, if only as a result of the already established tradition of tomb-robbing that was to dog the Egyptians' funerary customs throughout all history and outwit their best efforts to prevent it. Tomb-robbing was the price the Egyptians paid for equipping their dead, out of religious necessity, with all manner of this world's costliest goods. When the early ancient Egyptians realised that they had forfeited the preservation of their dead (and thus their chances of immortality, no less!) by furnishing them with bigger and better homes in which to spend eternity, they turned to the perfection of techniques of embalming and mummification that would restore the guarantee of survival if the physical body within the grave.

Early in the First Dynasty, in the years after 3100 BC, attempts were made at the better preservation of the bodies of the dead by means of wrapping them up in many layers of bandaging. An arm covered in costly bracelets was found in one of the royal tombs of the First Dynasty at Abydos which had formed part of a mummy wrapped in this way, and a Second Dynasty mummy from Sakkara shows the typical form of this sort of embalming with individually wrapped limbs and even fingers and toes. Apart from the wrapping, no additional efforts aimed at preservation seem to have been undertaken at this stage, though the use of gummed bandages enabled a quite lifelike moulded appearance to be maintained on the outside despite the decay within. In a Second Dynasty case, the body does appear to have been bundled up with raw natron salts – which were to play such an essential part much later on in the perfected version of Egyptian mummification. These salts were able, like the hot sand of previous times, to draw off the moisture that promoted the bacterial action of decay.

By the end of the Third Dynasty, at about 2700 BC, the technique of wrapping in gummed and resinated bandages was commonly in use for the mummification of the richer portion of society – no doubt the poorer went on being buried in simple desert graves throughout ancient Egyptian history. Inside the much-wrapped mummies of the day, a sort of internal combustion was initiated which annihilated everything but the bones of the deceased – at this time, the Egyptians could only hope to create a life-like external appearance that represented the living body, leaving the internal structure to deteriorate and disappear. The foot of the greatest king of the Third Dynasty, Djoser, who built the first of the pyramids (the Step Pyramid at Sakkara), was found wrapped and modelled into life-like appearance in the burial chamber under his striking monument: tomb-robbers had presumably ransacked the pyramid at an early date.

In the Fourth Dynasty, which was the great age of pyramid building when the Great Pyramid was raised for King Cheops at Giza, some new techniques were added to the art of mummification. The removal of most of the internal organs (the heart was left in place as the seat of consciousness) was practised now – these organs were embalmed separately and kept in their own separate containers near the body whose chances of preservation were greatly enhanced by the absence of the corruptible liver, intestines and stomach. The actual remains of the linen-wrapped viscera removed from a body for separate mummification (with traces of natron upon them) have been found in a box in one Fourth Dynasty burial shaft – in this case they belong to a very prominent person, none other than Queen Hetepheres the mother of King Cheops.

The appearance of names like Djoser, Hetepheres and Cheops attached to the human physical remains we are discussing marks a change in the character of our subject matter and our approach to it. Hitherto we have been looking at typical specimens of humanity and of the ancestors of humanity rather than at individuals in their own right (although of course every scrap of Australopithecus as much as every named mummy represents a real individual of the past). The Faces of the Past into which we have peered up to this point have served to chart the general evolution of mankind and the progress of human society from ape-like beginnings to the establishment of civilised life in settled and organised communities. From now on, even when we still do not know the names to go with the faces we unearth (as happens in the case of some Egyptian mummies as well as with the bog–bodies of north-west Europe), we shall be looking at individual men and women with individual stories to tell as much as they also reveal the general character of their times.

The Fourth Dynasty of ancient Egypt also saw, perhaps in conformity with the new technique of evisceration, a change from burial in a contracted to an extended position and some of the few surviving mummies of this time, which were still externally moulded with resinated bandages, have painted details too. From the next dynasty comes a striking mummy from the tomb of one Nefer at Sakkara (though the mummy may not represent Nefer himself) which was moulded with bandages soaked in plaster rather than resin and then painted to present a very life-like image of the dead man, with detailed nipples, genitals and face. With this Fifth Dynasty mummy we are not looking directly into a Face of the Past but rather at a stylised working-over of the dead man's real features beneath the moulded and plastered bandaging.

The first six dynasties of ancient Egypt constitute the 'Old Kingdom', which came to an end with internal strife and the collapse of central government in about 2250 BC. After 150 years of turmoil, strong centralised power was reasserted in Egypt for some two centuries before chaos crept up again. The period of stable government under the rulers of the Eleventh and Twelfth Dynasties was the 'Middle Kingdom'. During this period the art of mummification was revived in ancient Egypt, and some new techniques were tried out. One of them, seen to have been performed on the bodies of some Eleventh Dynasty princesses, involved the injection into the body, through the anus, of a fluid based on cedar oil and resembling turpentine, to help break down and then drain out the corrupted internal organs: the princesses' mummies do contain fragments of the decayed remains of the internal organs. The Middle Kingdom mummies are, in general, not very well preserved and one of them even contained the remains of a mouse and a lizard that had got into the mummy during the process of embalming. Insect larvae, it may be noted, are quite common in mummies of this and later periods. From a deposit of embalming materials of the Middle Kingdom comes an embalming table, consisting of a board raised up on four blocks of wood. It was in the Middle Kingdom, too, that mummies were first equipped with face-masks of plastered and moulded linen (called cartonnage) to go over their heads and shoulders, and also that mummies were first buried in human-shaped 'anthropoid' coffins.

The Middle Kingdom came to an end in the eighteenth century BC, not this time as the result of solely internal problems but with the assistance of an invasion of western Asiatic conquerors from Palestine and Syria, whose irruption into Egypt seems to have been part of an ancient world-wide process of turmoil and disturbance. The 'Hyksos' conquerors introduced the use of bronze into Egypt, along with light horse-chariots, at about the same time as the Bronze Age was being established all over the ancient world. In Mesopotamia, the law-giver King Hammurabi of Babylon had conquered all the lands between the Twin Rivers (except Assyria) a few decades before the Hyksos invaded Egypt. In Anatolia (in modern Turkey) Indo-European speakers, perhaps related to some elements in the Hyksos aristocracy of Palestine, had established the Hittite city states a couple of centuries before. At about the same time, the Indo-European-speaking (indeed early Greek-speaking) Mycenaeans had established their powers on mainland Greece and encountered the sea-power of the Minoans on Crete, around whose palaces like those at Knossos and Phaistos the first European cities were growing: these palaces were destroyed, perhaps by earthquakes, just before the Hyksos entered Egypt. At about the same time again, around 1750 BC, the cities of the Indus Valley civilisation were abandoned in north-west India in the face of the arrival of the Indo-European-speaking Aryans. In more distant China, the farmers in the Yellow River Valley were developing some distinctively Chinese traits, with the keeping of silk-worms and the furnishing of the graves of their ancestors with carved jade, and the divination of the future with the 'oracle bones' on which the first writing of Chinese characters appears. Even more remote from the ancient world of Egypt, in Peru in South America, farmers were domesticating the alpaca for wool, growing cotton and beginning to weave cloth.

The Egyptians, after about 250 years of foreign rule, threw out the Hyksos around 1560 BC and re-established native power in the shape of the 'New Kingdom'. The Egyptian rulers of the New Kingdom naturally wished to continue the practice of mummification and build themselves tombs suited to their newly restored power. Their tombs were cut in the desert wadis of the Theban Hills opposite the town and religious centre of Thebes where they seem to have originated (the modern Egyptian town of Luxor and the village of Karnak occupy the site of Thebes today). The pharaohs of the New Kingdom were buried in or near the Valley of the Kings for over 500 years and among their number are counted some of the most famous names of ancient Egypt. Their burials were subject to the attentions of tomb-robbers probably from the first, though with varying intensity of depredation from time to time. In the end, the priests of the Theban necropolis seem to have decided to collect together the surviving, if often battered mummies of the pharaohs in their charge and re-deposit them, after necessary bits of re-wrapping (for the robbers were no respecters of bejewelled persons or of any curses attached to them), in single group resting-places. It is from a great cache of mummies, discovered in 1881, and another similar but smaller collection found in 1898, that the mummies of most of the famous pharaohs come.

One mummy in the pharaonic collection now housed in Cairo's great museum, is that of Sekenenre who may, to judge by his vivid wounds (a dagger wound in the back of the neck, club and mace blows on the head and a wide axe gash in the forehead), have died fighting to expel the Hyksos from Egypt. His is not a very well-done mummy, in fact little more than scraps of skin and bone, and it would be consistent with the notion of his having been killed in battle to speculate that he received a hasty and make-shift mummification: when his case was recently opened to facilitate X-raying, his body was still found to give off an unpleasant odour of decomposition.

But the New Kingdom which Sekenenre and his immediate descendants ushered in saw the ultimate perfection of the techniques of mummification: there were repeated improvements of the process during the five centuries of the New Kingdom before another period of social collapse brought chaos to Egypt again. Afterwards, the art was again revived and in the fifth century BC, when the Greeks were a strong commercial presence in the land, the historian and travel-writer Herodotus was able to give a remarkably detailed, if sometimes inaccurate, account of the business of mummification as practised in his day. About 400 years later, the Sicilian Greek historian Diodorus, who travelled in Egypt in about 60 BC, wrote an essentially similar account of the process. Herodotus and Diodorus mention three methods of mummification. A very simple and cheap version of the process involved the drying of the body in natron salts, which is a sort of naturally occurring combination of washing soda and baking soda, to be found about 40 miles west of Cairo, that is able to draw off moisture and fluid very efficiently – precisely this sort of mummification seems to have been meted out in the case of some Middle Kingdom war-dead. A more expensive form of mummification called for the introduction of oil into the body through the rectum with a syringe to help dissolve away the internal organs – and we have seen that some Middle Kingdom mummies seem to have been made in this way. The most expensive style of mummification required evisceration and separate embalming of the viscera, the removal of the brain (which the Egyptians thought nothing of) through the nose and the seventy-day treatment in natron of the whole body before wrapping and delivery for the funeral. Experiments at the Manchester Museum have shown that natron salts, and not a natron solution as was previously divined from the text of Herodotus, are required to promote desiccation of the corpse (the Manchester researcher used rats) and it is now thought that the seventy days of Herodotus and Diodorus apply to the whole process of mummification and not just the natron treatment, as they supposed. Among other interesting details of their descriptions are the use of an 'Ethiopian stone' knife to make the evisceration cuts in the abdomen of the subject, the ritual chasing-off with stones and curses of the knife-wielder (Diodorus) and the method of removing the brain with hooks through the nose and infusions of dissolving oil (Herodotus). Diodorus says that the internal rinsing with palm wine and the packing with spices which they both mention made the corpse smell sweet. He adds that, in his time, some Egyptians liked to keep the bodies of their forefathers about them in fine chambers to look at 'the very lineament of their faces' as we are doing here. Of course, Diodorus was writing at a very late date in the history of ancient Egypt and of Egyptian mummification. None the less, the descriptions of Herodotus and Diodorus do fit very well with the archaeological evidence of the New Kingdom mummies themselves. The most elaborate and expensive form of mummification which they describe applies, with variations, to the New Kingdom royal mummies and later ones.

The collection of pharaonic mummies in the Cairo Museum has drawn crowds of visitors, almost certainly on account of the morbid fascination of these bodies from the fabled past. But the real interest of the mummies lies in the medical evidence they afford and the light they throw on ancient history. One of the earliest of the New Kingdom mummies, as we have seen, belongs to Sekenenre who may have died in action against the Hyksos. For an ancient Egyptian of approximately 30 years of age, Sekenenre had unusually good teeth. The ancient Egyptians did not suffer much from dental caries (which is associated with our modern high-sugar diet) but their food, especially their bread whose corn was ground with gritty grindstones, contained much abrasive material which frequently contributed to excessive tooth wear, with exposure of the pulp cavity and consequent decay and abscesses. Sekenenre's teeth were good and so were those of his grandson Amenophis I, who came to the throne in 1546 BC and started to extend Egypt's power both south into the Sudan and northwards towards Palestine. When his body was removed from its Cairo Museum case for X-ray examination in the 1970s, the investigating team were met with the sweet smell of the delphiniums wrapped in with the mummy.

Amenophis I was succeeded by his brother-in-law Tuthmosis I, whose son Tuthmosis II married his own half-sister Hatshepsut. After her husband's death (and a period of regency on behalf of a son by another wife, Tuthmosis III) Hatshepsut became King of Egypt in her own right, ruling between about 1495 and 1470 BC. By this time, the Babylonians and Assyrians were in power in southern and northern Mesopotamia respectively. The Egyptians themselves had reached the Euphrates and were in control of Palestine, holding some western

Asiatic peoples in bondage, among whom were perhaps some elements of the Hebrews. The Hittites were coming to dominate the more northerly part of the Middle East and were developing the large-scale use of iron for the first time. In the Aegean, the palace of Knossos on Crete was the focus for what was shortly to become Europe's largest township. On mainland Greece, the Greek-speaking Myceneans were trading with Cretan sea-empire; and in central and western Europe, the spread of bronze-working was being completed and such great monuments as the stone-alignments of Carnac in France and the spectacular stone-circle at Stonehenge in England had been built. In the East, the Shang Dynasty of China was based upon the towns of the Yellow River Valley – fine bronzes and silks were in production and the writing of the picture-script was being developed. In the Americas, farming was spreading to all parts of Central America and extending its range in South America to include parts of the Amazon basin, but there was still no metal working in America.

Hatshepsut's images in pharaonic art, where she is sometimes shown crowned and wearing the long plaited false-beard of kingly power, reveal a close resemblance to her brother/husband Tuthmosis II and her step-son/successor Tuthmosis III. There is a mummy in the Cairo collection that may be hers: a very well preserved body with a still characterful face and long dark hair, judged to be that of a person of between 35 and 45 years of age, which might suit identification with Hatshepsut, though no name was attached to this mummy by the priests who gathered the royal bodies together after the ravages of tomb-robbing. When Tuthmosis III finally succeeded Hatshepsut, he embarked on the extension and consolidation of the Egyptian New Kingdom Empire in a way that brought Egypt to the greatest peak of imperial glory, with its power stretching securely from the Sudan to the Euphrates. The mummy of Tuthmosis III shows him, like several conquerors who have come after him, to have been only about 5 ft (1.5 m) tall, but with pretty good teeth again for his time. There is a problem associated with the mummy ascribed to him in that, although he appears on historical grounds to have ruled for about 35 years, his mummy does not seem old enough, according to X-ray evidence, to have lived through this span of time; an interesting case where history, archaeology or X-ray evidence might leave something to be desired. His son, Amenophis II, continued his father's imperial policies from the

Fourth Cataract of the Nile to Syria. The mummy of his successor, Tuthmosis IV, reveals a very emaciated personage of about thirty years of age – interestingly, his ears are pierced.

The mummy of Amenophis III provides a wealth of medical evidence and information about the progress of techniques of mummification: he died at the age of about 50 years in perhaps 1367 BC, obese and bald and with a mouthful of very bad teeth coated in tartar and undermined by abscesses. The limbs of his mummy were packed with linen material to keep them filled out to their living proportions: this technique was to be elaborately developed later on. Amenophis III must have suffered grievously in the last years of his life and perhaps he neglected the duties of the great empire he ruled: it certainly looks as though his son and heir did, Amenophis IV, who initiated an astonishing revolution in ancient Egyptian affairs.

The mother of Amenophis IV was a lady called Tiy (attempts have been made to identify the mummy also ascribed to Hatshepsut with her, but it seems unlikely because of the mummy's apparent age at death) whose father and mother were Yuya and Thuya. Both their mummies survive in very vivid form, although they had been subjected to slightly different forms of mummification: Yuya's ethmoid bone was broken in the tell-tale way that indicates removal of the brain through the nose, while Thuya's nasal bones show no signs of breakage. Both had very bad teeth, with much wear and some loss of teeth and there is evidence of abscesses. Still Yuya's mummy, in particular, remains as an outstanding example of the characterful and life-like possibilities of Egyptian New Kingdom embalming. Their grandson Amenophis IV ruled for some years as co-regent alongside his ailing parent at a time when the priesthood of the great god Amun was coming to wield more and more influence and power. Perhaps as a political move to reduce the power of this priesthood and perhaps out of genuine religious inspiration (or more likely as a judicious combination of the two), Amenophis IV came to elevate another concept of deity, focussed on the disc of the sun called the 'Aten', above all other gods of Egypt. He renamed himself Akhenaten and moved his capital to a virgin site away from Thebes called Akhetaten, now known as el Amarna. Here Akhenaten devoted himself to his 'Amarna Heresy' which embraced not only religion and social practice but art as well. It seems likely that, years later when the Amarna Heresy had been utterly repudiated, Akhenaten's own mummy was dragged out of his tomb and destroyed to annihilate 'that criminal of Akhetaten' (as he is called in later texts) in this life and the next. Certainly no mummy of his has been found – but the more naturalistic art-style which Akhenaten fostered has bequeathed to us certain representations of him that show him as one grossly deformed, with feminine rounded thighs and belly and elongated face with lantern jaw. If these sculptures show him as he truly was (and it seems likely that they do) then Akhenaten was a pathological case of a sort that can even be identified with a certain degree of retrospective diagnostic reliability. It has been suggested that he suffered from 'Fröhlich's Syndrome', a condition leading to the sort of freakish features which his statues display and which also, incidentally, makes for infertility. All the more astonishing, then, to know that Akhenaten's queen (and the apparent mother of his children) was the gorgeous Nefertiti, whose famous portrait bust (and another less well-known, unfinished but perhaps even more appealing one) shows her to have been the most beautiful ancient Egyptian that we know. Her mummy is missing too.

The events and relationships of Akhenaten's reign have been the subject of much speculation and controversy. Ahkenaten himself has been variously interpreted as history's first real 'individual', as a noble prophet of monotheism (on the assumption that monotheism is a good thing), as a religious fanatic, as a political and social reformer, as an idler who let an empire go to pieces, as a homosexual partner of his final co-regent and perhaps brief-reigning successor Smenkhkare. It is certain that the Egyptians of later times execrated his memory and indeed tried very nearly successfully to erase memory of him altogether. Nefertiti has been variously seen as his inspiration, as the real power behind the throne, as his estranged queen and enemy: she has even been identified as the real person behind the name of Smenkhkare. There is a skeleton (mummy would be too strong a word for this decayed body) from a small tomb near the entrance of the Valley of the Kings that has been ascribed to Smenkhkare, though the coffin and tomb in which it was found never name the owner which they housed. This is the skeleton of a young man and his claims to be identified as Smenkhkare rest on his extremely close relationship, based on anatomy and even blood-grouping, with the occupant of another small tomb just across the way by the entrance of the Valley – one Tutankhaten.

It was Tutankhaten who came to the throne of the pharaohs after the brief reign of Smenkhkare and the two bodies from the two little tombs resemble each other so strongly as to be almost certainly those of full brothers, which makes the likelihood of Smenkhkare's having been a male individual, and not Nefertiti under another name, that much more plausible. Tutankhaten was perhaps only about ten years old when he became King at Amarna: his name shows that he had been brought up in the Heresy of Akhenaten whose half-brother he may have been; but within a few years his name was changed and Amarna was abandoned when the court moved back to Thebes and back, presumably, under the sway of the Amun priesthood who had seen off Akhenaten and his revolution. There must have been real powers-behind-the-throne during the few years of the young Tutankhamun's reign: we can identify them as the court official Ay, who succeeded Tutankhamun briefly, and the general Horemheb who succeeded Ay to some effect and glory for about 28 years. When Tutankhamun died at the age of about 19 years in 1343 BC, he was buried not in some great, deep-cut tomb in the Valley of the Kings, like many of his predecessors and successors, but in a small and probably hastily constructed tomb near the entrance of the Valley. True, his tomb was stuffed with precious things (and it has even been suggested that he was sent into the next world with goods originally intended for Akhenaten and Smenkhkare too) but his final resting place stands as much comparison with the great royal tombs of the Valley as a pre-fab does with Blenheim Palace. Perhaps it was the very paltriness of Tutankhamun's tomb that saved it for posterity and the devoted excavation of Howard Carter in the 1920s. The tomb was entered by robbers at an early date, within a very short time of the burial in fact, but the thieves were disturbed by the necropolis guards and the priests who were responsible for the royal tombs were able, albeit hastily and in a rough-and-ready way, to clear up the very limited damage done and reseal the tomb. The tomb was not entered again and, about 200 years later, the construction work of a later and grander tomb caused all traces of Tutankhamun's tomb to be submerged under stone-chippings and débris which effectively hid the tomb for a further 2000 years.

Tutankhamun slept on in his little tomb, surrounded by his riches, during all the time of the decline of ancient Egypt, the rise of Greek and Roman power, the arrival of Christianity, the collapse of the Roman Empire, the establishment of the European nation-states, the evolution of European capitalism, the expansion of the empire of an island in the far north-west – until an artist and draughtsman from that island of

Britain who had taken to Egyptology, called Howard Carter, engaged himself to an English aristocrat convalescing in a corner of his nation's empire after a car accident, called Lord Carnarvon, to locate and excavate the missing tomb of the now shadowy, long-dead pharaoh.

Carter began work on the tomb, which he had found after years of searching, in late 1922. He worked with painstaking patience on the outer rooms that held much of the treasure of Tutankhamun and was only ready to penetrate the burial chamber itself in 1925. Carter and his team had to open up the doors of three gilded wooden shrines, one inside another like a Russian doll, to reach the stone sarcophagus of Tutankhamun. Inside the sarcophagus they found three anthropoid coffins, even more like a Russian doll in shape and closeness of fit, the last of which was made of solid gold. All three carried idealised but recognisably individual portraits of the young pharaoh, as did a gold portrait-mask that covered the head and shoulders of the mummy. When the wrappings of the mummy itself were undone, the body of Tutankhamun was found badly blackened by the resins and unguents used in the course of mummification which seem to have set up a process of slow combustion within the tightly wrapped and multifariously encased mummy. It may even be that the mummy of Tutankhamun has fared badly alongside those of his contemporaries precisely because his body remained undisturbed by robbers inside his coffins, sarcophagus, shrines and tomb whereas those of his fellow pharaohs of the age were rudely evicted from their coffins and even pulled about by robbers to get at the precious objects wrapped up with them, which put a stop to their internal carbonisation. But, although Tutankhamun's mummy is not in the best of shape, his head retains a pathetic quality of vulnerability to set alongside the shining faces of his coffins and mask which show him in the full-cheeked bloom of youth. How he died we do not know – there is a trace of a possible blow on his face and it seems that the back of his skull was thin-boned in a way that might point to haemorrhage as the possible result of a blow, but we cannot tell what happened: it would not be unlikely that political expediency might have played a part in his exit. The man who carried out the unwrapping and anatomical examination of Tutankhamun's body lived, by the way, into his nineties and Howard Carter went on working on the Tutankhamun material until his death during the Second World War, but sadly Lord Carnarvon died from the after-effects of a mosquito bite, coming on top of his earlier road accident and not very robust health, in Cairo in 1923.

Chapter Nine

The Face of Ancient Egypt

1300–300 BC

The pharaoh Horemheb died towards the end of the fourteenth century BC. He arranged that his successor should be an old army comrade from the eastern Delta, one Ramesses, who ruled briefly as Ramesses I and whose mummy (like Horemheb's) has never been found. His son was Seti I, and his is one of the very best preserved mummies that have survived. Seti I regained much of the territory of the old empire that had been lost under Akhenaten and Tutankhamun and not fully restored under Horemheb. He built extensively in Egypt and in particular at Abydos, where numerous reliefs depict his profile in a manner that can be directly compared with his mummy: despite the fact that his head was detached from his body by graverobbers, his features have been described as the most life-like of all the royal mummies. His countenance is still blackened by the particular techniques of mummification in force at the time, but after him a lighter and more natural colouring was achieved.

Seti's son Ramesses II came to the throne in 1304 BC and campaigned extensively in western Asia, though not always with such total military success as his grandiose monuments (like Abu Simbel) and boastful inscriptions would have us believe. He died in 1237 BC and his light coloured mummy in Cairo Museum looks, even for a mummy, like the remains of a very old man: his teeth were badly worn and abscessed. The jaws and teeth of his son, Merneptah, show signs that some dental experts have tentatively attributed to the deliberate pulling-out of bad teeth: if this interpretation is correct, then here we have the only evidence for dental intervention among the Egyptian mummies. Ramesses III, at around 1200 BC, reigned over times of trouble for ancient Egypt that were related to turmoil and uncertainty across the whole ancient Mediterranean world. The Iron Age had been ushered in by the Hittites of Anatolia, and everywhere iron recommended itself by its relative abundance and ease of manufacture and its everyday superiority over bronze as the metal of tools and weapons. But Egypt had no natural resources of iron and economic difficulties accompanied the arrival of the Iron Age. In the east Mediterranean area, a wave of sea-piracy and coastal attacks had been set in motion by a loose and shifting confederacy of nationalities, among whom there were certainly some Greek elements. Ramesses III met and defeated some of these 'Sea Peoples' during his reign, but some of them were equally able to get a foothold on the Levantine coast and settle down to become the Philistines of Biblical history. At about 1200 BC again, the city of Troy was sacked by Greeks and the Israelites came to power under Joshua and his successors in much of the land of Canaan – some of them had probably been in bondage in the Egyptian Delta in earlier years. In the wealthy northern Canaanite city of Byblos, by modern Beirut, the Egyptian hieroglyphs were adapted to strictly alphabetical writing on papyrus sheets. The Sea Peoples played their part at about this time in toppling the Hittite Empire and sending survivors into Syria to become the Hittites of the Bible.

Despite the presence of Greek piratical elements in the Sea People coalitions, the trading power of the Mycenaean Greeks declined at about this same time and their cities were destroyed by the final influx of Greek speakers from the north in the shape of the Dorians. Relatives of these Greek speakers, the Latin speakers, were making their way from central Europe towards their final home on the Tiber during these furious years after 1200 BC. In central and north-west Europe the first beginnings of the European Iron Age were being manifested, with the identifiable presence of the Indo-European-speaking Celts. Further east in Mesopotamia, the Assyrians were rising to total power and the eclipse of the Babylonians; in India, the Aryans were spreading into the Ganges Valley, speaking an Indo-European language closely related to the language of the Celts, the Latins and the Greeks; while in China the Chou Dynasty was to overthrow the Shang in about 1100 BC. In Mexico the Olmec civilisation that underlies all the subsequent central American developments, was coming into being, while in Africa, south of Egypt, the Negro civilisation of Kush was developing on the Nubian Nile.

In Egypt a long line of Ramessides ruled until about 1100 BC. The mummy of Ramesses III is interesting as being the first to be equipped with artificial eyes of stone. Ramesses IV was given onions for eyes! Ramesses V of about 1160 BC may have had smallpox, to judge by the eruptions on the skin of his mummy, which was stuffed with sawdust to maintain a life-like fullness of form. The sister of the last of the Ramessides, the Eleventh, was married to Herihor, high-priest of Amun at Thebes in about 1100 BC. Her name was Nodjme and her mummy was packed all over to bring back a satisfying fullness to her limbs and face – her eyes were of stone and she was treated to false eyebrows (for the natron would have destroyed the originals). If it is not too much to say of such a grotesque and morbid production as a mummy, the face of Nodjme could be described as still beautiful. On the other hand, the packing technique could be taken too far and cause the face to burst open as has happened with another woman's mummy of the time. At about this same time, the practice was adopted of painting the mummies according to the 2000 year-old convention of Egyptian art – with red ochre for men and yellow ochre for women.

After 1100 BC, the power of Egypt in the ancient world declined, with internal division between the priests of Thebes and independent princes of the Delta. A Libyan dynasty took power in Egypt after about 950 BC and one of the rulers of this period, Sheshonk, campaigned with some success into Palestine, where – as the Shishak of the Bible – he sacked the temple of Solomon in about 930 BC. After him, a further decline into fragmented principalities occurred and, in the south, the kings of Kush won power so that they were able to frustrate native Egyptian attempts to reunify Egypt under an Egyptian king until after the Assyrians took control of the country for about a decade around 670 BC. All these years of trouble saw a decline in the standard of mummification, which became a matter of elaborately patterned external wrappings rather than a well-done job on the inside. There was, in fact, a sort of Gothic revival in Egypt in the sixth century BC, when a backward-looking enthusiasm for everything of the Egyptian past gripped the nation. At the same time, the Egyptian rulers were not above making alliances with Greek rulers and admitting whole Greek trading colonies into Delta bases. This was the beginning of a long Greek association with Egypt which issued in the rule of the Ptolemys, Greek kings of Egypt tracing their line back to one of the generals of Alexander the Great who conquered Egypt in 332 BC in the course of generally defeating the Persians who had held it on and off since 525 BC.

During these years after 900 BC of Egypt's decline and fall into the hands of foreign rulers, Mesopotamia saw the peak of Assyrian power pass into the Neo-Babylonian Empire of Nebuchadnezzar, to be succeeded by the empire of the Persians. The Assyrians conquered the northern Kingdom of Israel and the Babylonians the southern Kingdom of Judah, while the Persians returned the Jewish exiles to Jerusalem from their Babylonian captivity. In Greece, the cities revived from the dark ages that followed the collapse of Mycenaean power by about 800 BC and the foundations were laid for the flowering of Greek art and philosophy and various forms of political organisation – the Greeks defeated the Persians in Greece between 490 and 479 BC and the Macedonians went on to crush the Persian Empire under Alexander. In the west Mediterranean, the Canaanite Phoenicians from the Levant and the Greeks founded colonies, while in Italy the Etruscans flourished and ruled in Rome until the Romans expelled them in 509 BC to set up their own republic. The Aryans established the caste-based Hindu society in India, where Buddha taught until his death in about 480 BC; in the Far East working of iron was adopted in the feudal states of China and in the New World irrigation was developed in the valleys of Peru and the cultivation of plants spread to southern parts of North America.

During Ptolemaic times in Egypt, the practice of mummification was continued and the Greek rulers and many of their Greek subjects who lived in Egypt adopted ancient Egyptian customs of burial. Despite the maintenance of the elaborate methods of mummification detailed by Herodotus (who was in Egypt during the time when the Persians ruled there) and Diodorus (in Roman times), it seems that mummies were no longer produced to the standards of New Kingdom times. Very bodged and even fraudulent examples of mummification are known from these times, with bits and pieces missing and even the mix-up of parts from different individuals: all this suggests that the embalmers had become careless and cynical about their work and that, perhaps even more so than in the past, their workshops must have been gruesome and repellent places.

A mummy in the Manchester Museum collection which was unwrapped in the mid-1970s throws some very interesting light on the range of jobs which the late-period embalmers might undertake. Unwrapping of mummies in modern times had quite a vogue in the last century (when it was sometimes called 'unrolling') and was carried on until the turn of the century, when it went out of fashion. X-raying became a valuable technique for looking into mummies without the need to disturb them, but even the best of modern X-raying methods cannot reveal as much as actual unwrapping can do. Manchester Museum possessed an intriguing mummy, called simply 1770 after its museum number, which was thought to have come from a late nineteenth-century excavation at Hawara in Egypt and to be of probably Graeco-Roman date. X-rays had already shown that the body inside was probably that of a 13-year-old child, with the mysterious addition that the lower parts of the legs appeared to be missing. As the unwrapping proceeded, the mysteries of mummy no. 1770 deepened: the skull was found to be in a very fragmented state, with traces of red and blue paint on some of the cranial pieces; there was little sign of flesh upon the arm-bones and in the chest cavity. Two gold nipple covers were found which indicated that the body was female, for it was the ancient Egyptian tradition to equip the mummies of women in this way – on the other hand an artificial phallus, made of a roll of bandages, was found at the pelvis. The resins used in the course of mummification were found to have penetrated between the vertebrae and have reached the ends of long bones in the legs and arms. The broken-off ends of the legs turned out to have been fitted with dummy feet shaped out of reeds and mud, but shod with a fine pair of slippers. All in all, it seemed clear that the body of 1770 had been wrapped-up in an already pretty advanced state of decay, with very little flesh left on it. The unwrapping team, led by Dr Rosalie David of Manchester Museum, faced the possibilities that their young subject had been injured shortly before death with the loss of the lower limbs, or that the embalmers had done a spectacularly bad job on the mummification, or that the body had arrived for mummification in such a poor state of preservation (perhaps as the result of spending some time in the water – we know that persons drowned in the Nile were accorded honourable mummification at some periods of ancient Egyptian history) that the priests and embalmers in charge of mummification had been obliged to make the best of a bad job, and had even tried to cover all eventualities by equipping their subject with the attributes of both sexes. The anatomists on the unwrapping team concluded for their part that the remains of this adolescent belonged most probably to a girl.

Some of the problems associated with 1770 were resolved by subsequent radiocarbon dating: for it became clear that the body itself on the one hand and its wrapping and decorations on the other were of widely separated ages. The bandages were most probably manufactured and applied to the body between AD 100 and AD 400 (in other words probably during the period of Roman Egypt) whereas the body itself had lived between 1550 and 1250 BC, at some time in the New Kingdom. So it appears that the mummy of 1770 represents some young girl who died and was originally mummified during the New Kingdom but whose remains were re-discovered during the period of Roman rule and, though their sex was by now obscure, were judged important enough (probably they were identified as royal) to be accorded fresh mummification and re-burial.

As well as unwrapping the mummy of 1770, the Manchester team decided to make a reconstruction of the facial appearance of their subject in life. A similar exercise was undertaken with the mummies of two brothers of Middle Kingdom date that were in the Manchester collection. The method employed was essentially similar to the one used by Gerasimov in reconstructing the living faces of our remote human ancestors: both procedures go back to the work of Kollman and Buchly in Germany at the end of the last century, when they established the depth-range (according to age and sex) of soft tissue at 26 different points on the face. When the reconstruction work was done on the face of one of the two brothers, the result (objectively achieved by the application of the soft tissue depth-rules to his skull) revealed a face recognisably similar to a wooden portrait bust found in his tomb. In the case of 1770, the first task was to re-assemble the fragments of bone making up her skull: the traces of blue and red paint on them were, of course, explained by the fact that the body was in a near-skeletal state when it was re-wrapped by the Roman-period embalmers. The re-assembly of 1770's skull disclosed a rather lop-sided face containing several unworn teeth, indicative of ill-health and the life of an invalid who had been fed on special food. From the state of tooth-eruption in the jaws, the age of 1770 was judged to be approximately 13 or 14, confirming the opinion generated by the earlier X-rays. When the skull fragments were re-assembled, the Manchester team made a flexible mould from them and cast a replica on which to build up modelling clay according to the depth-rules for the soft tissues that had clothed the bare bones of 1770 some 3500 years ago. Despite the scrupulously objective technique employed in 'fleshing-out' the face of 1770, the team soon felt that their subject was taking on a distinctive character of her own. Photographs chart the stages of reconstruction as the modelling clay was built up on the face and details like eyes and hair were added. The Face of the Past which emerged from the completed reconstruction of the skull of this young girl of the Nile Valley of some 1400 years BC, is one of a slightly invalid vulnerability. It is both sad and curious to reflect on the short life of this unnamed but evidently high-born child who must have lived a sickly life until her death and first mummification on the threshold of womanhood, only to be almost certainly disturbed by tomb-robbers at some time after her burial (perhaps early on, for tomb-robbing was a steady occupation in ancient Egypt, or perhaps during one of the periods of economic trouble when tomb-robbing was rife) and then, more than 1000 years after her early death, to have her broken remains mummified anew and, 2000 more years after that, to be scientifically dissected more than 2500 miles from her home! But the researchers of the Manchester Museum have succeeded, with their investigation and reconstruction, in giving 1770 something of what the ancient Egyptians wished for themselves after death – a sort of immortality, but sadly without our knowing her all-important name.

If we compare the face of 1770 as reconstructed by the medical artists on the Manchester Mummy team with the faces we encounter in

Egyptian art, we see at once that the art does not really encompass the depiction of anything like the living face of the New Kingdom girl. Egyptian art (like all representational art, as a matter of fact) is stylised and conventional in its depiction of the world, including the world of human beings. It is notorious that Egyptian art was both highly stylised and conventional, to our eyes, and very conservative in that it went on representing things in much the same way over the whole 3000 years or so of its span. There is much truth in this popular perception of the ancient Egyptian art, although in fact it did evolve and change during its long career, maturing early on in Old Kingdom times into a classical form. This was modified during the New Kingdom in particular (and quite revolutionised during Akhenaten's Heresy in the direction of an idiosyncratic naturalism), and then fell into decadence during the late period with some brave attempts to revive the Old Kingdom style before the rather hideous and mechanical productions of Ptolemaic and Roman times finally ruined it.

What is apt to look to us like extreme stylisation in Egyptian art from the first was dictated by the different religious outlook and requirements of the ancient Egyptian artists as compared with our own habits of mind. Eternity was at the heart of the Egyptian temper: the ancient Egyptians looked for the eternal essentials of things rather than wanted to capture fleeting moments of individual lives! The world of nature and the animals recommended itself to them because it appeared so unchanging alongside the world of human history. The religious requirements of the Egyptian view of the afterlife, in which the body and the various spiritual components of every dead person needed to be preserved for ever the same, reinforced the necessity to make art embody the eternal essentials of man. And so the ancient Egyptian artists devised and adopted in perpetuity a convention of sculpture, bas-relief and painting which sought to show simultaneously all the essential attributes of man in a timeless combination. In bas-relief and painting especially, convention required that, for instance, both arms and the chest be shown semi-frontally while the rest of the body was in profile (but one leg often in front of the other, so that both were present). Statues and pictures, moreover, showed their subjects in an idealised state of young maturity, so that their everlasting vigour was assured. The aged Ramesses II, for example, is never shown in his triumphant statuary as an old man. The Egyptian art is, in fact, an art in the perfect tense and never in the present imperfect: its subjects are never really shown in the midst of performing a transient action as though caught by the camera; they are shown, even when for example they appear to be striding forward in that pose so characteristic of some of the best of their statuary, in a perfected state that eternally possesses the act of striding.

But within their strict conventionality of expression, the ancient Egyptians were capable of the most vivid humanity, which goes far beyond the art of their contemporaries among the first civilisations: you have only to compare their Old Kingdom art with the works of the Sumerians and their Mesopotamian successors to recognise in the latter a lack of humanistic quality. But look at the statues of the Old Kingdom Prince Rahotep (with his little 'Latin Lover' moustache) and his plump wife to see the humanity of Egyptian art!

The Egyptians were among the first art-masters of the Greeks: the Greeks always retained a reverence for the immensely older and well-established civilisation of Egypt, tending to credit the Egyptians with even more depth of innovation in human history that they deserved. There is little doubt that the Archaic Greek prototypes of classical statuary, the so-called 'Kouroi' figures of gods like Apollo from the early sixth century BC, with their stiff hands-by-the-sides pose, were derived in their general stance (if not in the expression on their faces) from Egyptian models. The Greeks of classical times and later, and their Roman imitators, were to advance the humanity of representational art after the native Egyptian art had fallen into its terminal decadence.

Chapter Ten

The Graeco-Roman World

1500 BC–AD 500

At about the same time as the young 1770 was enduring her short life, and in that same Bronze Age period which saw the New Kingdom of ancient Egypt, the Mycenaean Greeks were flourishing on mainland Greece. In fact, the first Greek-speaking people probably moved into Greece at about 1900 BC, at a time when speakers of related Indo-European languages were moving into India as the Aryans, Italy as the Latins, and central and western Europe as the Celts. By 1600 BC, these Greek speakers in Greece had founded fortress cities like Mycenae, after which they are named, and entered into commercial relations with the Minoan Empire based on Crete, where the splendid palaces of Knossos and Phaistos were located. After about 1400 BC, the power of the

Minoans declined and the Mycenaeans, who had taken on board much of the Minoan culture, stepped into Cretan shoes as traders and colonisers in the east Mediterannean.

The Mycenaeans, with their bronze tools and weapons, were a warrior people, ruled by local kings and a warrior-aristocracy. It is these warriors and their way of life whose bright memory Homer celebrates in the *Iliad* and *Odyssey.* The works of Homer in some ways took the place of a Bible in the lives and thinking of the later classical Greeks and Romans and it was Homer's works, more than 2500 years after the traditionally blind bard composed them, that inspired a nineteenth-century German amateur archaeologist to rediscover the lost world of the Mycenaeans. Nineteenth-century scholarship was largely persuaded that Homer's works were pure fiction, but Heinrich Schliemann regarded them as something very close to 'Gospel Truth', and set out to dig up Homer's heroes in the citadels Homer placed them in. Schliemann started his career of discovery in 1871 at what he believed to be the site of ancient Troy near the Dardanelles. He discovered not one long-lost city in the mound of Hissarlik, but a whole series of them: he truly did discover the site of Troy, but we know he misidentified the city of Homer's King Priam in the stratigraphical sequence of ruins within the mound.

After his Trojan triumph, Schliemann turned to the impressive ruins of Mycenae in Greece, which Homer gave as the citadel of the powerful King Agamemnon who led the Greeks against Troy. Scliemann ignored the huge open and empty tombs outside the walled citadel with its great lion-gate, and dug inside, where he found five deep grave-shafts just inside the wall. At the bottom of these grave-shafts, he found the burials of men and women and their rich grave-goods. Two male bodies wore golden masks over their faces. Schliemann, of course, concluded that he had found the bodies of Agamemnon and his companions slain by his adulterous queen Clytemnestra on his return from Troy. And Schliemann fixed upon the finer and grander of the gold face-masks and declared 'I have gazed upon the face of Agamemnon'. We now believe that the real historical events which underlie Homer's account of the Trojan War (which he may have composed, on the basis of orally transmitted bards' songs and poems, more than four centuries after the event) took place in about 1200 BC or shortly thereafter, at the time when the Sea Peoples – among whom there were Greek elements – were ravaging the coastal cities of the east Mediterranean. The grave-shafts of Mycenae, on the other hand, belong to about 1500 BC and are contemporary with the early years of the New Kingdom in Egypt. Indeed the tradition of furnishing the dead kings of Mycenae with gold face-masks is probably derived from the Egyptian example (of which a surviving instance is to be seen in the case of Tutankhamun). Schliemann had not really looked into the face of Homer's Agamemnon but he had stumbled on the first spectacular evidence of the existence of a brilliant European civilisation contemporary with that of New Kingdom Egypt. The 'Mask of Agamemnon', as it is still called, is on display in the great museum of Athens, along with those of other Mycenaean kings and kinglings. It and the others are nothing like life – or death – masks, for they are extremely stylised in the details of eyes, ears, beards and moustaches, though they are individually rather characterful at the same time: one of them portrays a genially round-faced man, while 'Agamemnon's' is a seemingly strong and severe face. Although we are hardly looking into real individual Faces of the Past when we gaze at these Mycenaean face-masks, we are looking at the bearded and mustachioed expression of the fierce warrior people whose coming to power in Greece laid the foundation of the civilisation of classical times that underlies all European culture to this day.

The island civilisation of the Minoans based on Crete with whom the Mycenaeans were closely associated in trade and cultural exchange, declined after about 1400 BC, very possibly as a result of the devastating volcanic explosion of the Minoan island of Thera, called Santorini today, which sent a great destructive tidal wave and hail of ash all over the Minoan world. Faint memories of this cataclysmic event, perhaps transmitted through Egyptian records, may lie behind the legend of Atlantis of which the classical Greek philosopher Plato wrote a thousand years later. The Mycenaeans seized the opportunity created by the collapse of the Minoan power to move into the trading network formerly operated by Crete: there is evidence to suggest that they may have done so even before the destruction of Minoan civilisation, for some of the records of the Cretan palace of Knossos were already written in Greek before the end came.

The Mycenaeans never quite equalled the civilised splendours of the Minoans they supplanted in power on Crete and the islands of the east Mediterranean. They lived more like Anglo-Saxon or Viking chiefs with rich and costly possessions among the squalors of their towns, whereas the Minoans had displayed – at least in their rulers' palaces – a degree of refinement (including plumbing and sewers) that would not be seen again till Roman times. Carousing in their timber halls and listening to the epic tales told by their bards to entertain and flatter them, the Mycenaean warlords must have greatly resembled the Anglo-Saxon chieftains of 2000 years on.

When the Sea Peoples brought strife and upset to the east Mediterranean world at about 1200 BC, though there were Greek speaking kinsmen among these pirate hordes, the Mycenaean trade routes were wrecked and Mycenaean power declined. Perhaps the attack on Troy, which Homer remembered as being led by the King of Mycenae itself, was motivated by a desire to end the Trojan city's monopoly of the route to the Black Sea.

In the vacuum of lost Mycenaean power, fresh waves of Greek speaking peoples moved south into Greece from east-central Europe, in the historical shape of the barbarous Dorians and a Dark Age fell upon the Greek world.

The Dorians probably arrived in Greece, via the Balkans, from east-central Europe in about 1100 BC, already equipped with Iron Age technology. They spoke a dialect of the same language as the Bronze Age Mycenaeans whom they conquered, displaced and often, no doubt, married. Some of those whom they displaced sailed away to Aegean islands and to the west coast of Asia Minor, now part of Turkey. It was in these 'Ionian' communities that the bardic memories of the Mycenaean heroes were kept alive over three or four centuries and transmitted into the hands of the poetic genius (or geniuses, for some scholars believe that more than one poet worked on the epics) of Homer. After about 800 BC, the Greek cities revived from their Dark Age slumbers and the character of classical Greece emerged. The period of this emergence, from 800 to 500 BC, is called the Archaic, when among the many traits of the mature classical civilisation the Greeks developed, initially out of Egyptian prototypes as we have seen, their own distinctive sculpture of the human form and face. Their first 'Kouroi' statues are stiff and formalised according to the Egyptian model, but often their faces wear the extraordinary wide-mouthed smile that so well expresses a certain sort of serenely self-satisfied and benevolent beatitude, the 'Archaic Smile'.

The Greeks had been trading with Egypt since at least the seventh century BC, and in the sixth the pharaoh Amasis had consigned the Delta town of Naucratis close to the Mediterranean, to them as their commercial base: it was while the Persians were in control of Egypt after 525 BC that Herodotus visited it and learned about the methods of mummification. Herodotus, who was born in Halicarnassus in Asia Minor (Ionia) in about 480 BC, also travelled east from his home to visit the tribes who lived along the northern shore of the Black Sea in and around the Crimea: the Scythians, who fascinated him hugely. He gives a very full account of them, including their splendidly barbaric funeral rites which involved the internal cleaning of the corpses with frankinsense (akin to parts of the Egyptian practice of mummification), human sacrifice and the mourners' inhaling of hemp fumes! Much of Herodotus' account has been corroborated by archaeological excavation and spectacularly so in the case of the tombs of some nomads closely related to his Scythians found in the Altai Mountains. Here, at Pazyryk, the tombs of these nomads have remained permanently frozen, and their perishable contents, which would have disappeared without the frost, have survived: among the contents are felts and carpets, (which sometimes show both Persian and Chinese influences), clothes and boots and the bodies of a chieftain and his wife torn from their coffins by tomb-robbers. The woman is revealed as a European type, whose long black hair had been cut off and interred in a case of its own. Her brain and internal organs have been replaced by plant materials. The man's body is of Mongol features, and his skull carries two axe-gashes and the signs of scalping. His body was tattooed in early life (for the pigment had penetrated muscle before he grew fat) and even if we cannot in his case look into another Face of the Past, at least we can regard one of the earliest examples of tattooing known in the archaeological record, on the body of this nomad chieftain who died around 2500 years ago, at just about the time when Herodotus was describing the life of the Scythian barbarians north of the Black Sea.

Until the later sixth century BC, the Greek sculptors were still bound by the stiff style they had derived from the Egyptians, but by the mid fifth century BC, such formalities were left behind and the Greeks were producing lively and realistic, if still idealised, portrayals of the figures of gods and men. Such was the head of Pericles, who guided democratic Athens through the first years of its long war with the warrior-aristocracy of Sparta – a war which ended with the destruction of the Athenian Empire but not, thanks in part to Spartan magnanimity, with the destruction of Athens itself which survived to become the university town of the later Greek and the Roman world (Plato's Academy flourished in Athens for some 900 years, until suppressed by Christians).

The city-states of Greece were never united into a single political structure until the Greek speaking Macedonians, under their King Philip and his son Alexander (who had been tutored by Plato's successor Aristotle), defeated Athens and her allies in 338 BC. After Philip's murder, Alexander carried out his father's plan to invade and conquer Persia, which had become the great power of the east Mediterranean world and, though it had not succeeded in overrunning Greece, had occupied much of the Middle East including Egypt. Alexander liberated Ionia from Persian rule, then Syria and Palestine and then Egypt, where he was proclaimed Pharaoh in 332 BC. For over 11 years, Alexander's conquests took him – and Greek civilisation – through the whole Persian Empire as far as north India, where actual Greek rule obtained for a period under Alexander and then again a century later. A school of religious art arose in an area that is now in India and Pakistan that was based on Greek models: it lasted for about seven centuries and very possibly embraced the first representations of Buddha (who had died in about 480 BC) in human form, looking like an Indian version of Apollo. But Alexander died young, in 323 BC at the early age of 32, and then his empire was divided among his generals: Ptolemy got Egypt and founded a line of 'Ptolemaic' Greek rulers, of whom Cleopatra was the last. Like Egypt, the whole of the ancient Middle East became Hellenised, with Greek as the universal language of commerce and culture and, of course, Greek civilisation was modified and 'orientalised' in various locally different ways.

In Egypt the descendants of Alexander's general put on, for official purposes, the garb of the native rulers and were depicted, in rather decadent and deformed reliefs, as the heirs of the pharaohs. But above all in the coastal city of Alexandria built between the sea and Lake Mareotis on the eastern edge of the Delta, the Greeks in Egypt practised an art and a way of depicting human beings that was as far removed from the official art of the Ptolemaic pharaohs as it could be. Alexandria was founded and named for Alexander himself in 332 BC: it became the greatest port of the ancient world and, like all great ports, it was intensely cosmopolitan in character. Here not just Greek and Egyptian ideas mingled, but there came together all the social, religious and artistic traditions of the ancient world through the medium of the universality of the Greek language. The culture that emerged in Alexandria and became the stock-in-trade of the whole ancient world is called 'Hellenism', after the Greeks' own word for what was generically Greek (our word comes from the Romans' 'Graecus'). Hellenistic art advanced the portrayal of the human face and figure beyond the classically idealised expression of the old mainland Greek artists, towards an anecdotal realism that encompassed new subject matter drawn from the fleeting moments of everyday life: look at the face of a drunken old woman from a piece of Hellenistic sculpture (known to us in a Roman copy)!

The Romans became a power in the east Mediterranean world after their final crushing victory in 146 BC over their great rival city-state, Carthage, which was by origin a North African colony of the old Canaanite-Phoenician traders of the Levant. With Carthage destroyed, the Romans went on to involve themselves in Greek and Middle Eastern affairs until, during the first century BC, they had absorbed most of Alexander's old empire in the region. The hill-top town of Rome had been founded, according to its own tradition, in 753 BC. The Romans were descendants of peasant farmers who had come into northern Italy from central Europe by at least 1000 BC, distant cousins of the Dorians who entered Greece at about the same time, and speaking an Indo-European language related to Greek and the languages of the Celts in Central and Western Europe and the Aryans in India (and to the languages of most of the peoples, excepting the Basques, Hungarians and Turks, living in all of Europe from Portugal to the Urals today, as well as the Iranians and Hindus). The Romans were at first completely overshadowed by the Etruscans living in their wealthy Tuscan cities to the north of Rome and, to a remoter extent, by the great Greek colonies in Italy to the south. The art of the Etruscans and of the Greek city-states in southern Italy served as the example for the Romans during the centuries of their emergence as a great power, but it was above all to the Hellenistic art of the east Mediterranean of the second and first centuries BC that Roman sculpture and painting owed its inspiration, at a time when Rome was sending its sons to Athens for their education (or recruiting Greek tutors to come and teach in Rome) and looting the collections of Greece and the Hellenistic world for its own adornment.

After the Romans cast off their Etruscan kings in about 500 BC, they turned to a republican form of government which saw them through to the conquest of all Italy by about 300 BC, the defeat of Carthage in the Punic Wars which ended in 146 BC and the annexation of much of Alexander's old empire by 30 BC. By this time Caesar, who often acted impatiently of republican forms and niceties, had conquered Gaul and made two forays into Britain, bringing the Romans into close contact with the Celtic peoples of Western Europe and the Germanic tribes of the north. Caesar's great-nephew Octavian gained eventual control of the entire Roman world and its conquests in 31 BC, ruling under the title 'Augustus' as the first of the Roman emperors.

During the years of the Roman Republic, up until about 30 BC, the wide world beyond the reach of the Graeco-Roman consciousness (which could extend, as we have seen as far as the steppes of Russia and the north of India) continued to develop along its own lines. In China, Confucius taught his code of social life in the late sixth and early fifth centuries BC while the Taoists advocated withdrawal from the affairs of the world, which became increasingly hazardous in China during the period of the 'Warrior States' between about 400 and 220 BC. After 200 BC the Han Dynasty re-organised China according to Confucian principles and it was during this period that certain burials were made in China in which the bodies of their dead occupants have survived down to our time in a quite outstandingly good state of preservation. According to inscriptions found in one of these tombs, the deceased was a man named Sui who died on 13 May 177 BC. We know that he was a provincial official of the Han Dynasty, and about 60 years old when he died. The Chinese archaeologists who unearthed Sui in 1975 ascribe his excellent state of preservation to the 'fluid' in his tomb: Sui's body was submerged in an anaerobic liquid which was retained by the structure of his tomb throughout more than 2000 years; it does not appear that he was deliberately embalmed. At all events, his body was preserved in a very fresh state: supple and soft, resilient to the touch, and not a bit shrunken; his brain was intact in his skull, his eyeballs in their sockets, his eardrums in his ears – the ivory docket with his name on it was found in his gullet. His skin retains a reddish-yellow colouring, suggestive – if not of life – of recent decease: his is probably the most natural-looking Face of the Distant Past that we can look upon today.

Another of the Chinese bodies is that of the Marquise of Dai, a woman of approximately 50 years of age whose cadaver, after the injection of preserving fluid, remains full and springy like Sui's. Her clothed body was laid in the tomb in an arrangement of coffins resembling Tutankhamun's, packed under layers of charcoal and clay which account for the microbe-free preservation of her remains. These Chinese tombs often contain models of boats and figures of human beings and animals in very much the same way as the ancient Egyptian ones did, to serve the dead in the next life. China did not, it should be noted, remain forever quite outside the orbit of Rome: after about AD 75 the silk route was opened up to exchange Chinese woven cloth for Roman glass and wool.

During the last few centuries BC, the Negros of the Upper Nile and the Niger rivers, who had developed an Iron Age culture based on cattle-rearing, are generally thought to have spread south into the rest of the continent, taking their Bantu language with them. They reached the Cape, where Bushmen and Hottentot natives had lived for millennia, at about the same time as the Dutch settlers of the seventeenth century AD. Meanwhile, in the Americas, the course of civilisation initiated by the Olmecs in Mexico after 1300 BC was advanced with the development of a complex calendar-system at about the time of the end of the Roman Republic and the rise to civilisation of the Zapotecs in the Oaxaca Valley of Mexico after 300 BC. In South America, the Moche civilisation was building canals and aqueducts and great platforms of sun-dried clay along the northern coast of Peru, in the interests of a highly-organised militaristic pattern of society. The art of the Americas, Central and South, possesses a distinctive quality from the first which distinguishes it from any of the arts of the Old World. Though some aspects of the natural living appearance of its Amerindian creators are apparent in the human representations of this art, there is a frequently grotesque look about it that prevents its opening up for us any direct window on the American Faces of the Past, however much it tells us about the psychological disposition and artistic outlook of its makers! The Peruvian Incas of much later times than the last years of the pre-Christian era did produce some natural mummies, bound and shrivelled bodies from dry-cave burials, that show us a few dead faces from their past, but it is to the sculpture, pottery and painting of the American peoples that we are restricted when we want to confront them in their antiquity.

The Romans who took over much of the empire of Alexander, including Egypt with its cosmopolitan city of Alexandria, came into direct contact with the Hellenistic art that was being produced all over the Greek and Greek-influenced ancient world in the last two centuries BC. The Romans naturally went on to take this Hellenistic art as their own model when their wealth and power inspired them to furnish their public places and their homes with sculpture and painting. They also looted a great deal of antique Greek art back to Rome and copied its styles in their own productions. The Roman market for art was immense, both for the domestic enrichment of their homes and the grandiose decoration of their public monuments, especially after the overthrow of the old austere republic. The portrait busts of the increasingly vainglorious emperors, their courtiers and women, over four centuries of power chart the changing fashions of the times (beards come and go, for example, and feminine hairstyles are elaborated) and constitute for us a vivid gallery of the personalities of Roman imperial history. The funerary art of tombstones extends our insight into the look of the inhabitants of the Roman empire over wider sections of society.

In Egypt, the Greek and Roman middle-classes of merchants and imperial administrators often liked to be buried in the Egyptian way, as mummies. The latter-day mummies of Egypt were not very well-made on the inside and most of the embalmers' efforts were put into the production of impressively packaged mummies with intricately patterned bandaging (sometimes studded for further effect). These mummies of the Graeco-Roman period of ancient Egypt often carry a feature unknown in earlier days, and quite out of character with the native Egyptian style of human depiction: portraits of the deceased painted on wooden boards and affixed to the mummy over the head of the corpse inside by binding their edges into the wrapping of the bandages. It seems that these paintings were done during the lifetime of the deceased and were kept at home until the deaths of their subjects, so they do not necessarily represent their owners at the time of death but more probably in the prime of life. Certainly, they are very vivid portraits of the men and women of Graeco-Roman Egypt, often hauntingly gazing out at us with a direct look in their eyes: these poignant Faces of the Past seem to be looking at us as much as we are regarding them. The portrait-boards may have belonged originally to larger and fuller figures of their subjects which were cut-down to size on their deaths and incorporated into the wrapping of their mummies.

There is another source of information which gives us a picture of the lives and deaths of some of the citizens of the Roman Empire in the early years AD. This source belongs to Italy itself, south of Rome and close by the city of Naples, in the shadow of the ominous volcano called Vesuvius, where the Roman town of Pompeii once flourished. The first rumblings of trouble for Pompeii and Herculaneum, and the other places that were later to be destroyed with them, came in February of AD 62, with earthquake damage which the citizens of these towns made good over the next few years. It was on 24 August AD 79 that the volcano Vesuvius blew with the loss of thousands of lives and the burial of the communities around it under ash and lava which have preserved much of their structure to this day to yield such spectacular archaeological discoveries since excavations began in the seventeenth century. History and archaeological investigation come together at Pompeii and Herculaneum, for an eyewitness account of the terrible events of the three days of lava-flow and ash-fall in August of AD 79 was given by a Latin writer in letters he sent to the Roman historian Cornelius Tacitus. At Herculaneum hot mud-lava engulfed the town and many, if not most, of its citizens escaped the immediate danger, though very recent excavations at the now-inland site of Herculaneum's old beach suggest that a great many people died there before they could be taken off to safety. At Pompeii, pumice and volcanic ash rained down upon the town and the people there were killed in large numbers by falling stones, by fumes, by the suffocating ash and, of course, by their own panic.

Pompeii became a city of the dead, with more than 2000 corpses of people trapped in the streets or in their own homes by the all-enveloping hot ashes which built up over them and eventually hardened over their bodies, preserving their shapes and features like moulds. Inside these cooled and hardened ash-moulds, the bodies decayed away leaving only scraps of bone behind: when archaeologists came upon these human-shaped cavities in the hard ash-deposits of the sites, they devised a method of introducing plaster into them to make casts of the living (in fact, dying) appearance of the men, women, children and pets of Pompeii. What we see at Pompeii are not the physical remains of human beings and (in some cases) animals but casts made from the natural moulds created by the consolidation of the volcanic ash which fell all over them. The casts present a life-size and substantial but boldly-sketched image of their original subjects: faces, in these cases, are not finely-detailed but roughly expressive; and what they mostly express, inevitably, is the agony of their ends. Look at the face of a man, against the far wall in a roomful of contorted bodies, raising himself despairingly on one elbow to see the Face of the Past of Pompeii, AD 79. Among the bodies that have been 'recovered' by the casting technique are those of a beggar trying to escape with his sack of alms; two chained-up gladiators who could not escape their barracks along with their fellows; and a magnificently bejewelled woman who was, in some circumstances or other, paying a visit to the gladiatorial barracks. We see a man pulling a goat; a tied-up dog; a woman with a baby and two small children; a pregnant woman; children dead in their playroom; a woman who appears to have died serenely and men who died raging against their fate. There are priests of the Temple of Isis who had tried to preserve their holy things; also a family where the children were embracing and the parents holding hands and a funeral party who had died trapped at the tomb of the dead they were mourning. In one place of death, someone – perhaps a Jew or perhaps a Christian – had scrawled on the wall the two words 'Sodom, Gomorrah'.

The art of the Roman Empire, during its five centuries of production, was of course subject to stylistic evolution: by the late third century AD, something of the severe formality of the later Byzantine art of the Christian Roman Emperors of Constantinople was becoming evident. The carved figures of the Tetrarchs of about AD 300 (the emperor Diocletian and his co-regents) show this stylisation very clearly: their proportions, their stance and the expression on their faces (they all have the same one) are a world away from the art of classical Greece and Rome.

In about AD 370, a little temple of the Egyptian goddess Isis belonging to the port of Corinth on the Greek mainland was destroyed by earthquake. Still crated up (and never installed, because of the destruction of the temple) were some glass panel pictures from Egypt: when one of them was recovered from the sea a few years ago, it was found to depict the poet Homer whose *Odyssey* and *Iliad* fulfilled for the pagan world something of the rôle of the Bible in Christian times. The two epic poems and their mythological author (his existence in anything like the form attributed to him by tradition must be regarded as being as dubious as that of Moses) were treated with deep reverence. The late fourth-century picture of Homer from Corinth shows him bearded and strong of countenance and irresistibly recalls the later Byzantine depictions of Christ. It seems likely that the Christian artists of the Byzantine Empire modelled their faces of Christ upon the severe visage of the most revered figure of the pagan ancient world: Homer as represented by the artists of the late Roman Empire.

Chapter Eleven

The European Barbarians

200 BC–AD 400

During the last century of the Roman republic, when Rome's power was spreading around the entire Mediterranean world and taking over the old east Mediterranean empire of Alexander and his heirs, there lived in north-west Europe various tribes of Iron Age people whom the Greeks and Romans numbered, despite their commonly inherited Indo-European languages, among the 'Barbarians' or 'gibberers' of outlandish tongues. The technology of iron had reached Europe, from the home of its first widespread application after 1500 BC as a state-monopoly of the Hittite kings of Anatolia, shortly after 1000 BC. Previously, the peoples of north-west Europe had been living in their Bronze Age – often in circumstances of considerable wealth and social complexity.

The builders of the final, spectacular phase of Stonehenge (whose ruins most people today identify with Stonehenge per se) were Bronze Age chieftains of a rich commercially-based society on Salisbury Plain in England, trading between Ireland and central Europe at about 1500 BC. At much the same time, a wealthy Bronze Age culture was flourishing in Scandinavia which has left behind some striking cult-objects like the trumpet 'lurs' (still to be seen on portions of Lurpak butter!) and the gold disc of the sun pulled on its chariot by bronze horses that was found at Trundholm in Denmark. Exceptional circumstances of preservation in oak tree-trunk coffins have bequeathed us several bodies dressed in the clothes of the Danish and north German Bronze Age of approximately 3500 years ago. Though the heads of hair of these Bronze Age corpses have sometimes survived very well, their flesh has almost completely vanished and their bare skulls grimace at us in place of their living faces. But their clothes have survived wondrously well, to reveal that the men of these times wore long tunics wrapped around the body and fastened over the shoulder, with a belt around the waist and a cap on the head, while the women sometimes wore – along with enveloping cloaks – a sort of prototype of the miniskirt made of woolen cords.

The trousers that so exotically caught the eye of Greek and Roman writers (whose own togas led them to regard the breeches of the Barbarians with disdain) are absent among the Bronze Age tree-trunk burials of north-west Europe: they belong to the Iron Age Celts and Germans of the last few centuries BC, and are probably owed to the influence of the horse-riding nomads of the Steppes whom Herodotus had encountered as the Scythians around the northern shores of the Black Sea.

The first use of iron in central Europe can be traced to the area of Bavaria and Upper Austria, where an ancient cemetery in the Salzkammergut has given its name to the first phase of the European Iron Age: Hallstatt, about ten miles south-east of Salzburg. The prehistoric communities of this place amassed their wealth as a result of their salt-mining, and salt of course figures prominently in the place-names of the region. The earliest evidence of iron-working in this region belongs to about 900 BC, at much the same time as the initiation of the Iron Age of northern Italy and about 200 years after the first appearance of regular iron-use in Greece: the Iron Age seems to have reached the western coast of Europe and the British Isles in the eighth century BC. The second phase of the European Iron Age, after 500 BC, is named after a French site called La Tène and this place is identified with the emergence of the typically Celtic art style, with its emphasis upon curvilinear and abstract designs rather than upon naturalistic representation of the world. The early Celtic art does contain examples of the representation of the human form and face but their incorporation into the abstract compositions of the style leads frequently to a grotesque exaggeration. Later on, Celtic art achieved a 'plastic style' which included representations of the human form such as the striking sculpted head from Czechoslovakia which looks for all the world, with its turned-up moustache and eyebrows, like – as Sir Mortimer Wheeler once put it, comparing its features with his own on a television programme – 'a Celtic brigadier'.

The Celts of central and west Europe embarked upon a series of migrations in the late fifth and fourth centuries BC, which brought them into contact with the emerging power of Rome (and included the sacking of Rome itself in about 390 BC) and took them as far as Spain, Greece and even Asia Minor, in the latter of which some of them settled to become the 'foolish Galatians' of the letters of St Paul. The Romans defeated the Celts in northern Italy around the turn of the third century BC, and – having seen off the threat of the Carthaginian Hannibal at about the same time – were able to extend their conquest to the Celtic Gauls who lived north of the Alps in present-day France, where they sacked the Gaulish stronghold at Entremont in Provence in 123 BC: a deed recorded by that same Diodorus who described the Egyptians' methods of mummification. The end of the second century BC saw the Roman pacification of southern Gaul: but the tribes who lived to the north of these southern Gauls continued to give trouble. The Cimbri and Teutones (who have given us the word 'Teutonic') rampaged south from their homeland in modern Jutland (Denmark) and Schleswig-Holstein (Germany) and were only finally defeated by Roman armies in northern Italy in 101 BC. The Romans went on, under Caesar, to conquer all Gaul by the middle of the first century BC, and to make a couple of exploratory raids on southern Britain and cross the Rhine on expeditions against Germanic tribes harassing their Celtic cousins in Gaul. During the reign of the first Roman Emperor, Augustus, the brothers Drusus and Tiberius (who eventually became the successor of Augustus) were able to carry Roman conquests into southern Germany and reach the Danube. The various German tribes outside the Roman Empire continued to give trouble. One of them, under a king the Romans called Arminius – his name was Herman in his own tongue – annihilated a Roman army in AD 9, but Herman was later assassinated and the ensuing stability in Germany permitted the Romans (under Claudius, the son of Drusus) to invade Britain in AD 43.

The great Roman historian, Cornelius Tacitus, wrote an account of the Germans against whom the armies of his country campaigned, and included in his account the best reports he could obtain of the way of life of German peoples and tribes beyond the Roman frontier 'only recently become known, whom war has revealed to us'. In a good many particulars, his account has been astonishingly corroborated by the discovery since the last century of the actual physical remains of some Germanic tribesmen of north-west Europe's Roman Iron Age: the period covering the last few centuries BC and first few centuries AD when the Iron Age barbarians of north-west Europe were heavily influenced by association with the Roman world.

Tacitus tells us such details about the Germanic tribes as that they threw cowards and unwarlike men (as well as sexual transgressors) into muddy bogs with wooden hurdles fixed over them; that adulteresses had their heads shaved before being driven out of their villages; that they wore the skins of wild beasts; that their meals were simple – wild fruit, fresh game and curdled milk; that the tribe of Suebi (whom we should call Swabians, after their German home in Schwaben) took their hair over to one side of their heads and knotted it low down, and that they indulged in human sacrifice.

Since the early years of the last century (and probably before, but not recorded), hundreds of preserved bodies have been discovered in north-west Europe, with a concentration in the north German land of Schleswig-Holstein and the Danish province of Jutland. Preserved bodies have also been found in Holland and Britain and other parts of Germany and Scandinavia, but they have rarely survived beyond the immediate time of their discovery and were frequently, until very recently when archaeology has gained greater public prominence, thought to be remains of tinkers, peddlars, recent murder-victims or drunks and treated to a quick, Christian reburial in hallowed ground where they are beyond the reach of further research. The Danish (about 170 bodies) and north German (about 70) examples are the only ones to have been recovered in sufficiently large numbers for a pattern of dating to be established for them: where they are dateable, they mostly belong to a period between about 200 BC and AD 500, in other words the Roman Iron Age. Some of them no doubt found their way into the bogs which preserved their bodies by accident, getting lost on dark nights and in storms and fogs, or by foul play at the hands of robbers or personal enemies. But such explanations, as we shall see, do not appear to suit the majority of well-preserved cases: most of the bog-bodies represent the victims of human sacrifice or capital punishment in the Roman Iron Age. Their date is vouched for by, in some cases, the application of the radiocarbon dating method and by the stratigraphy of the peat bogs in which they have been found, where a particular layer of sphagnum moss is known to have formed during the Iron Age. The bodies owe their preservation to the physical characteristics of the bogs: all air was excluded from contact with the bodies and the bog-water was saturated with acids capable of preserving (in fact tanning) the flesh of the bodies buried in the peat – though too much acidity has destroyed the bones of some of the bog-bodies, leaving only the well-preserved envelope of skin looking as though a road-roller has gone over it.

It has been the post-war discoveries of bog-bodies in Germany and Denmark that have excited the greatest public interest. In 1948, near Osterby in Schleswig-Holstein, the head of a man of about 50 or 60 years of age was found in a peat-cutting, wrapped in a sewn cape of roedeer skin. This head had been struck off its body with some sharp weapon that cleft the second cervical vertebra in two; there was not in this case much skin left adhering to the skull, but the hair was well-preserved – gone reddish with the acids of the bog, but once blond and flecked with grey, it had been gathered at the right side of the head in a low-tied knot, just as Tacitus reported was the fashion among the Suebi tribe in particular.

Another post-war German bog-body find confirms the general accuracy of Tacitus' description of the north European barbarians of his time: in 1952, the body of a girl was found at Windeby in Schleswig-Holstein, in – as so often is the case – an ancient peat-cutting which the Iron Age people had used to inter their victim. That the Windeby girl (who was about 14 years old at the time of her death) was a victim of the fierce code of the Iron Age Germanic tribes is made clear by the state of her well-preserved body. Her blond hair had been shaved close to the skull on the left side of her head (it was about 2 in (5 cm) long on the right) in much the way that Tacitus describes the penalty among these peoples for adultery; and although there were no signs of violence upon her body (she was probably drowned in the bog) the girl was blindfolded and a birch-branch and stone lay across her body to hold her down. According to the evidence of pollen percentages and types of pottery found in the same old peat-cutting in which she lay, the Windeby girl had lived and died in the first century AD. X-rays of her remains disclosed that the brain had been well-preserved inside her skull, which is not uncommon with the bog-bodies, since the human brain contains chemical substances that are not soluble in water, and the Windeby girl's brain was duly revealed for investigation: here we are looking not just at the face but at the thinking organ of the past!

The Danish bog-bodies have been turning up in Jutland for at least two centuries past: one of them, a woman's, found in the last century, was mistakenly identified with the queen of the Viking Erik Blood-axe and rests to this day in a chapel. This body represents a woman of about 50 years of age and there is reason in her case to think that she was still alive when she was staked out in the bog, under wooden crooks with transverse branches across her chest and belly – for one of her knees appears to have swelled up under the restraint of a willow crook. The provision of oak stakes and willow crooks and branches to hold down the victims of the Iron Age bogs is a not uncommon feature of the bog-body burials: in one case a veritable cage of branches was recorded, recalling the remarks of Tacitus about hurdles fixed over the victims.

The area around Borre Fen in Himmerland, in Denmark, has been the scene of some important archaeological discoveries since the end of the nineteenth century. There is an early Iron Age hill-fort nearby and a later Iron Age village has been identified there, in which the men and women whose bodies have been found in the bog may well have lived. In 1891 the very important Gundestrup Bowl was found in Borre Fen: this silver-gilt cauldron lay in pieces and was restored to a bowl of about 27 in (0.7 m) in diameter; its 13 elaborately embossed panels carry images of some of the gods and goddesses of pagan Europe on the outside and on the inside, scenes of religious ceremony. The style of the imagery on the Gundestrup Bowl is judged to place its manufacture in the Celtic rather than Germanic world, probably in Gaul during the first century BC. Perhaps the Cimbri took it back to Jutland as loot during their southward rampage at the end of the second century; later on it was placed in the bog as a sacrificial offering. The scenes upon the Gundestrup Bowl do not, then, necessarily represent the specific gods and religious ceremonies of the Germanic tribes, indeed they can hardly do so very closely. At the same time, it seems certain that the images on the bowl [illegible] nevertheless mean a great deal to the people who acquired it, treasured it and offered it to their gods in the bog. There was, after all, a common linguistic and cultural background shared between the Germans and the Celts (and, for that matter, the rest of the Indo-European-speaking world), and, beyond that, a continuity of some religious beliefs across the whole ancient world – the Gundestrup Bowl's scenes are thought to encompass Celtic, classical and oriental elements. And there is one particular scene upon the bowl that does seem to reflect the practice of human sacrifice that was among the religious traditions of the Iron Age Germanic tribes: sacrificing (perhaps by drowning) an up-ended man over what may be a great cauldron.

In 1946, the first of the bog-bodies from Borre Fen was discovered: the man lay in what was evidently an ancient hole in a disused, Iron Age peat-digging, on a layer which enabled the find to be dated in the first century BC. He had been put into this hole in a sitting position, but had later been compressed by the build-up of the peat; his skin was black and leathery and, after his disinterment, his left eye opened up by itself to reveal a yellowish eye-ball. The good state of preservation of his hands made it possible to guess that he had not been used to hard manual work in life; his stomach contained the remains of a vegetable meal; the branch of a birch tree lay across his body and there was a hemp rope round his neck, indicative of hanging or strangling. A year later, the body of a woman with a crushed skull was found in the same fen, lying naked on a sheet of birch bark with her upper part covered by a blanket and a shawl; there were some bones of an infant lying alongside and the right leg of the woman had been fractured some 4 in (10 cm) above the knee; perhaps this woman had transgressed against the sexual code of her tribe in some way. The next year, in 1948, a plump woman's body was found in Borre Fen; her face was crushed and the back of her head had been scalped.

The most famous and most affecting of all the bog-bodies of the Iron Age was found in 1950 in Jutland at a site about six miles west of Silkeborg, in a boggy area called Bjaeldskovdal. The man whose body came out of the bog in May of that year is known to the world as 'Tollund Man', and his head is on show today in Silkeborg Museum.

The peat of Bjaeldskovdal has always been highly prized and peat was dug there in the Iron Age: ancient cuttings and spades of that time have been discovered there. The first bog-body to be found in the bog was discovered in 1927, but it was quickly lost again under collapsing soil before there was time to examine it. In 1938, the body of a woman – known as the Elling Woman – was found there, hanged or strangled and wrapped in sheepskin, with a long plait of hair from the crown of her head that was very probably gathered in the Swabian knot of Tacitus and some of the other bog-bodies. About 60 yd (55 m) away from the place where the Elling Woman was found, the Tollund Man was recovered 12 years later from under eight feet of peat. So well preserved was the Tollund Man, and so vivid the expression of his face, that he looked to the peat-cutters who found him like the victim of a recent murder: the Silkeborg police were called and they were fortunately aware of the archaeological possibilities of such bodies, alerting the academic authorities at Aarhus University.

The Tollund Man's head, before the peat block in which he was found was removed to Copenhagen for full investigation, was to the west, facing south. He was wearing a sewn skin cap, fastened over his head with a hide thong; there was a hide belt about his waist and a leather rope noose around his neck; his hair (reddened by the bog acid) was close-cropped and his chin clean-shaven, except that a stubble had grown on it. His seemingly serene expression is the most striking and indeed affecting thing about his face: it is hard to believe that he was hanged in some, to us, barbarous rite nearly 2000 years ago; he almost looks as though he might be only sleeping. The eminent Danish archaeologist P.V. Glob, who worked on the Tollund Man as on many other of the bog-body finds, has called his 'the best preserved head to have survived from antiquity in any part of the world'. The rest of his body, dark-tanned all over like his head by the acid of the bog, was very well-preserved, too, with his legs drawn up and his arms bent in front of his face, though some patches of decomposition had occurred on his left side which had been uppermost in the peat. Sadly, because the costly techniques of museum-conservation were not perfected at the time of the Tollund Man's discovery, only his head, a foot and thumb now survive to be displayed in Silkeborg Museum.

The Tollund Man was between 30 and 40 at the time when he died, and his body on excavation measured about 5 ft 4 in (1.6 m) tall, but it was judged that he had shrunk somewhat in the peat (the skin lay in shrunken folds on his bones). That he had been hanged was confirmed on close inspection by the furrow on the skin of his neck at the sides and under the chin, left by the braided leather rope round his neck. His body was subjected to an intensive autopsy, which showed that all his internal organs were intact: in his stomach and intestines there were found the remains of his last meal, taken between about 12 to 24 hours before death. There were no traces of meat in this last meal, only the remains of plants and seeds: evidently it had been a sort of porridge of barley, linseed and other seeds (thirty different kinds in all). The composition of the meal has been taken to indicate that the Tollund Man was killed in winter or early spring, and it is open to speculation that his death may have formed part of a seasonal ritual to do with ensuring the fertility of the crops on which these Iron Age farmers depended for their subsistence. It is strange to think that we probably know at what time of the year the Tollund Man died, but we do not know which year it was in the range from about 260 to 180 BC indicated by radiocarbon dating (which puts the Tollund Man among the earliest of the Iron Age bog-bodies that have been well dated). The actual meal consumed by the Tollund Man on the last day of his life was reconstructed for a BBC television programme of the 1950s and received badly by Professor Glyn Daniel, who later described it as 'tasteless rubbish', and Sir Mortimer Wheeler, who jested that its awfulness might have accounted for the Tollund Man's 'suicide, because he could stand his wife's cooking no longer'.

In fact, the Tollund Man was almost certainly sacrificed to the gods of his northern tribe in some forgotten religious ritual. Medico-legal experts who examined his head have expressed the opinion that his eyes and mouth were probably closed by his survivors, indicating some care on their part and he seems to have been deposited in the bog in a comfortable sort of sleeping position. Certainly, the calm expression on the face of the dark-tanned and slightly shrunken head of the Tollund Man is likely to inspire us with the hope that he met his end in some state of faith in the gods to whom he was sacrificed, whatever their credentials. And what man was unable to do with all his embalming techniques for the truly life-like preservation of his dead in ancient Egypt, nature had done with the acids of the Bjaeldskovdal bog for the Tollund Man.

About 11 miles to the east of the site of the Tollund Man's discovery in 1950, there was found two years later an altogether less tranquilly countenanced bog-body: that of the Grauballe Man. Whether as a result of the grimmer circumstances of his death (his throat was cut from ear to ear) or the pressure of the peat in which he was preserved, the Grauballe Man's twisted body and agonised grimace are worlds apart from the serenity of Tollund Man. Pollen analysis and radiocarbon dating concur in placing his death at some time between AD 200 and AD 400, at least four centuries later than Tollund Man's. Like Tollund Man, Grauballe Man's skin is dark and his hair red from the action of the acids of the bog: he was deposited naked in his peat-cutting. His last meal, eaten immediately before death, substantially resembles that of Tollund Man in its vegetarian character and the small bones and hair that might indicate meat are probably owed to the presence of rodents in the grain-store from which he was fed. Again, the composition of his meal, lacking both in berries and in greenstuffs, suggests death in winter or early spring and reinforces the notion of sacrifice to gods who had to do with fertility. The coincidence of the nature of Tollund and Grauballe Man's last meals further suggests that the meal itself was of a ceremonially prescribed kind. Although the Grauballe Man's teeth were heavily worn and showed some decay, his well-preserved hands and feet – with clear prints and whorls – show little sign of his ever having done any heavy work. Taken with similar evidence from some other bog-bodies, the possibility is raised that the preferred sacrifices of these northern barbarians may have been selected and privileged persons, perhaps priests themselves. The Grauballe Man's cut throat, be it noted, could not have been self-inflicted but was positioned so that it must have been done by someone else.

Some of the bog-bodies of the north European Iron Age were buried naked, and some were minimally clothed: the evidence for clothing includes skin caps, open-work bonnets, capes, skirts and scarves. From this evidence, it seems that the women usually wore long garments not unlike the simplest of the Greek women's clothes, the peplos, while men often went about clad only in a short cape: a surprising fact which is consistent with the depictions of some Germanic prisoners on Roman military monuments. Leggings and shoes were added to this kit and there was probably a long wrap-around outer cloak for bad weather. While a tight 'body-stocking' of linen may also have been part of men's costume (some of the figures on the Gundestrup Bowl appear to be wearing such a garment) the evidence for the wearing of trousers and sleeved shirts seems to belong to a late phase of the Roman Iron Age in north-western Europe: the Schloss Gottorp Museum in Schleswig has examples of such clothing from the bogs. Trousers and shirts were to go on to become the standard, basic male costume of Europe in modern times.

The bog-bodies of the Roman Iron Age of the northern, barbarian tribes beyond the frontier of the Roman Empire, spanning the centuries around the BC–AD line, represent the remains of the ancestors of the Germanic and Scandinavian peoples who came to power in north-west Europe after the end of the old Roman Empire. Many of these northerly barbarians saw service as mercenaries in the Roman army and, when they retired back to their native hearths, they often took their military equipment with them as prized possessions (they also, no doubt, stole some of their Roman paraphernalia in the course of border skirmishes with the Romans). In the centuries after about AD 500 when the Saxons, Angles, and Jutes and other Germanic tribes were taking on their historical character, and the more northerly Scandinavian tribes were developing their own variants of the common language background they all shared, the shining example of Roman military style and equipment was remembered, copied and developed: the striking helmets of the Anglo-Saxons (like the one found at Sutton Hoo in East Anglia) and the Vikings are descended from the fine parade-helmets of the later Roman army.

While the Romans were administering (and, increasingly, defending) their huge empire from the Atlantic to the Middle East, and the northern barbarians were worshipping their obscure and sinister gods, the ebb and flow of political power and the advancement of the store of human technological achievement were proceeding around the globe. In Persia, the power of the Parthians, whose empire had rivalled Rome's, was in decline, to be succeeded by the Sassanid Dynasty after AD 226, with their splendid capital at Ctesiphon, where the religion of Zoroastrianism (which emphasised the dual powers of Good and Evil in the Universe) flourished. In India the ancient Aryan religion of Hinduism was elaborated at centres of Sanskrit learning (Sanskrit is the classical language of India, with strong affinities with Greek and Latin and the rest of the Indo-European tongues) and Buddhism entered a phase of sectarianism – in northern India after about AD 460, the same Huns who brought trouble to the Roman Empire were creating disturbance. In China, the Han Dynasty fell in AD 220, and barbarian invasions brought disruption in the north after AD 300, but the use of paper for writing was pioneered in these years and the dim beginnings of printing were initiated. The Iron Age arrived in Japan, where the Shinto religion of ancestor-worship was developed and the Yamato clan who came to power adopted many of the traits of Chinese language and culture. In Mexico the Olmecs and Zapotecs continued to flourish and pyramids were built (more than 2000 years after the Egyptians had finally abandoned them as a major form of monument) at Teotihuacan, which thrived as a huge city until about AD 600, while the Mayas developed their own hieroglyphic writing system and devised a complex calender of greater accuracy than anything known in the Old World at the time; in Peru the Moche were building their aqueducts and great platforms of sun-dried brick.

In AD 284, the Roman Emperor Diocletian divided his empire into a western and an eastern wing, acknowledging an inherent split that rested upon the Latin character of the western and the Hellenistic and oriental character of the eastern halves of his huge and financially troubled domain. Diocletian, a man of humble origins, took to affecting the hauteurs of an oriental despot and so changed the style of both the pagan and Christian emperors and the popes who succeeded him. The Roman Empire was thoroughly infected with oriental attitudes and religious beliefs by this time: cults like the ones of Mithras from Persia, Isis from Egypt and Jesus from Palestine were rife. The Christian religion, which had been developed in a form that recommended itself to the Roman world by St Paul in the AD 50s, was persecuted savagely by Diocletian. One hundred and fifty years later, and without giving up any of the pretensions of his title except the claim to personal divinity, the Emperor Constantine became a sort of Christian; in AD 380 Theodosius made Christianity the state-religion. When the western half of the old Roman Empire collapsed in the face of barbarian onslaughts from East Europe and the north, after AD 476, the emperors in Constantinople at the eastern end of the Mediterranean continued to rule over Christian Orthodoxy for another thousand years, before they were ousted by the Turks. In the West, the barbarians who had fragmented the western empire into smaller kingdoms (and paved the way for the nation-states of modern times) became Christians too, and the popes in Rome –

remote from Constantinople and unburdened by any single, powerful political authority close to home – were able gradually to rival the emperors of the East in their self-aggrandisements. Even the inauguration of the Holy Roman Empire in the West, when the Frankish King Charlemagne was crowned (by the Pope, in Rome) on Christmas Day of AD 800, did not diminish the standing of the popes in Rome: the entire Roman Empire had been donated to them by no less than the Emperor Constantine himself (according to a papal forgery of the eighth century AD)!

Chapter Twelve

The Face on the 'Turin Shroud'

Probably about AD 1350

All Europe, East and West, and the northern part of the Mediterranean world (together with parts of its eastern end) were Christian by AD 1000: even the first Russian states around Kiev and Novgorod were converted to Christianity by that time; but it should be remembered that another religion, whose origins were closely connected to Christianity's, had threatened to overwhelm much of European Christianity for a time – Islam.

Islam is the most recently propagated of the monotheistic religions with roots that go back to the Hebrew monotheism of the Old Testament: indeed, it has been suggested that Mohammed, who died in AD 632 of the Christian calendar, was as much influenced by desert heretics of Christianity as he was by Judaism. Both Islam and Christianity are developments, one way and another, of the Hebrew religion that emerged from a generally Semitic background in Palestine in the centuries before and after 1000 BC.

Hebrew history is uncertain before that time and the Bible itself is an inevitably unreliable guide to history, because it was brought to its present form at a much later time than the events with which it deals, and at the behest of political and religious motives that reflect the pressures of the sixth century BC (and later) rather than realities of a thousand and more years before. The situation is very much as it is with the writings of Homer, except that the time-gap between alleged events and their writing-up is often much greater and the religious and political considerations much more insistent. And so the stories of Abraham and his relatives, for example, must be set aside from any objective historical account of the emergence of the Hebrews and their religion. The story of Moses and his people's escape from Egypt is similarly dubious (although it may belong to an earlier stage of Biblical composition than the narratives about Abraham) and may well reflect the adverse experience of the Jews deported to exile in Babylon by Nebuchadnezzar after the sack of Jerusalem in 586 BC. What seems clear is that the Israelites came to power in the old land of Canaan at about 1200 BC, perhaps under a military leader called Joshua who captured Jericho and other Canaanite cities but perhaps, equally, as the result of a social and religious revolution within the fabric of Canaanite society. In fact, it is very unlikely that the strict monotheism of later times, which declared that only one god existed, was in force at this early date: rather, the Israelites may have practised a 'henotheism', which adhered to one particular god as their own god and left other peoples to struggle on with their own (no doubt inferior) deities.

Under David and especially his son Solomon, the Israelites prospered after 1000 BC, seeing off the threat of the coastally-based Philistines (whose ancestry included some of the Sea-Peoples who had brought trouble to Egypt and the whole east Mediterranean world a couple of centuries before) and trading widely with their neighbours in the oriental world. Solomon built the First Temple in Jerusalem, which was wrecked by Nebuchadnezzar nearly four centuries later, and his 700 wives included an Egyptian princess. After Solomon, the kingdom was divided between the northerly Israel which fell to the Assyrians in 722

BC, and southerly Judah which fell to the Babylonians (who had overcome the Assyrian Empire) in 586 BC. It was during the Babylonian Exile that the Jews first developed their mature concept of themselves as a chosen people, of the one God of the universe who had chosen them and of their uniquely God-centred history from Abraham, through Moses to all their troubles in the world. When Cyrus the Persian liberated the Jews in Babylon, which he conquered in 539 BC, and allowed them to go home to Jerusalem, they saw him as the instrument of God's will. In Jerusalem, the returned exiles set about the rebuilding of the Temple and, later on, the walls of their city.

When Alexander the Great defeated and appropriated the Persian Empire, the Jews came into the orbit of Hellenism and, while some of them went on determinedly adhering to the strict code of religious beliefs and practices they had begun to perfect in their years of adversity, others among the Jews of the ancient world (living not just in Palestine, but in Hellenistic cities in Syria and Egypt too) were heavily influenced by the thought and philosophy of the Greeks. From the mixture of Jewish and Hellenistic philosophies and faiths came some of the strands that make up new patterns of religious belief in Christian times.

Judaism itself, whose 'official' history and character is witnessed in the Bible, seems to have been more diverse than the Bible would have us believe. The careers of the prophets, like Elijah, with their bouts of wizardry against the false gods of foreigners, and their generally extravagant style of preaching, point to the possible presence of a variety of religious expressions within and around orthodoxy; while communities like the Dead Sea Essenes of the last century BC and first part of the first century AD reveal a wealth of sectarianism within 'normative' Judaism that tells us a lot about the religious complexity of the years between the Old Testament and the New. There are foreshadowings of some Christian belief and practice and even of some of the very sayings attributed by the Gospels to Jesus in the Scrolls of the Qumran Community found in caves above their monastery by the Dead Sea. It has even been possible to suggest that the career and sayings of Jesus may reflect, not the activities of a real person living in the first third of the first century AD, but a distillation of a thousand strands of Jewish, Hellenistic and generally ancient-world religion (including Mesopotamian and Egyptian elements) made originally among some Jewish sectarians, developed by the religious genius of St Paul and, after the

Romans' destruction of Jerusalem in AD 70, spread through the Roman world among gentile as well as Jewish communities. Be that as it may, it is certain to say that the life of Jesus himself is still beyond the reach of objective historical investigation, being charted only in the writings of his followers and lacking any independent corroboration.

All religions have their shrines and relics: Buddha's tooth, the Tomb of David, the Prophet's Cloak, Lenin's cadaver, and so on. To the faithful of their particular cult, these things are objects of devotion: to the sceptical, they are as often as not mere curiosities and perhaps distasteful ones at that. The Christian religion has been well-endowed with relics since the pioneering days of the Emperor Constantine's mother, Helena, whose collection at Constantinople came to include material ranging from the Crown of Thorns to the measuring rod by which Noah's Ark was laid down. Many Christian relics were destroyed in Paris by the rationalists of the French Revolution, and probably the best-known and the most controversial Christian relic of today is the Sindone or 'Turin Shroud'. Is this piece of old cloth the veritable Shroud of Christ, crucified in Palestine in the first century AD, or a brilliant forgery of say mediaevel times in Europe, or perhaps an accidental product of some other time and place? And what – quite apart from questions of faith, which presumably has deeper grounds than the validity or non-validity of an object like this – might be the significance of the 'Shroud' for archaeology, history and art-history?

What we can fairly safely assume to have been the same cloth as the one now in Turin Cathedral was being exhibited in the town of Lirey in north France, some 90 miles from Paris, by the late 1350s AD as the burial cloth, or shroud, of Jesus. Such exhibitions of relics, not always as ambitious as to belong to Jesus himself but to the many saints of the church, were far from uncommon at the time, but a French bishop named Henri de Poitiers was sceptical about the Lirey shroud and set up an investigation into its credentials. The unfavourable result of his investigations caused the exhibitions at Lirey to be stopped, but they were started up again during the time of Henri's successor as bishop of Troyes, Pierre d'Arcis, who wrote to the Pope at Avignon with history's first mention of this 'Shroud': 'sometime since in this diocese of Troyes, the dean of . . . Lirey, falsely and deceitfully, being consumed with the passion of avarice, and not from any motive of devotion but only of gain, procured for his church a certain cloth cunningly painted . . . upon which . . . was depicted the two-fold image of one man, that is to say the back and the front . . . said to be the actual shroud in which our Saviour Jesus Christ was enfolded in the tomb, and upon which the whole likeness of the Saviour has remained thus impressed together with the wounds He bore'. Pierre goes on to tell the Pope that his predecessor Henri de Poitiers 'eventually, after diligent enquiry and examination, discovered the fraud and how the said cloth had been cunningly painted, the truth being attested by the artist who had painted it, to wit, that it was a work of human skill and not miraculously wrought or bestowed'. Not, you might think, the most enthusiastic endorsement at the outset of its career of a relic that was to retain its popular appeal and attract scientific attentions 600 years later on!

The Pope decreed that, whilst the Lirey Cloth might be in future exhibited, it must be put forward only as a 'representation' of the Shroud of Jesus, and not the real thing. The Church has never since taken up an official position on the authenticity of the cloth, but in the six centuries since its first exhibitions, enthusiasts for the 'Shroud' have, of course, forgotten the Pope's pronouncement as to its being only a representation.

That we may be reasonably confident that the Lirey Cloth and the 'Turin Shroud' are one and the same is indicated by the substantial agreement between the account of it given by Pierre d'Arcis and what we see today and by what we know of its subsequent history, during which time it has been fairly frequently depicted by artists in conformity with its present appearance. All this does not entirely rule out, of course, subsequent modification of the image or even, at a stretch, a substitution. The cloth is actually about 14 ft (4 m) long and 3 ft 6 in (1 m) wide: it is made from two pieces of linen of almost the same length, one just over 3 ft (90 cm) wide and the other strip about 4 in (10 cm); both pieces are woven in three-to-one herringbone twill, and are stitched together. People who have seen it describe the cloth as yellow with age and the image as a faint brownish mark. The image depicts, as the bishop stated, the back and front of a man: he is full-length, with his hands folded across the genitals which are not visible. He appears sturdy, of perhaps above average height for any period at about 6 ft (1.8 m), wears a beard and shows shoulder-length hair. The cloth bears a number of stains, apparently caused by bleeding, which show carmine in a good light. The head has a series of marks, which have been said to have been caused by blood flowing from small puncture wounds. The body and legs are covered with clusters of stains: there are stains proceeding from the wrists and the right foot; another stain extends from the right side of the chest. All have been associated with Christ's Passion: wounds inflicted by the Crown of Thorns, the scourging, the nailing to the Cross, the spear-thrust of the soldiers; it has even been claimed that bruises are detectable on the shoulder of the image caused by carrying the Cross. So great is the coincidence of the details upon the image of the 'Shroud' and the Gospels' account of the death of Jesus that one can be sure at once that the Lirey Cloth is not an accidental product of some other decease but, if it is not the genuine Shroud of Christ, must be a deliberately detailed forgery. If the cloth could be shown to date to about AD 30, then it *might* be the Shroud of Christ (though it might still be a forgery, of course): if not, then it is certainly a forgery.

The history of the 'Turin Shroud' is now fairly well documented from the time of its first exhibitions, but not before. After its controversial exhibitions in the time of bishop Henri, who died in 1370, and bishop Pierre who protested to the Pope about it in 1389, it remained in the Lirey church until the middle of the next century, when it was sold to the House of Savoy who built a chapel specially to contain it at Chambéry in 1502. That chapel caught fire in 1532 and the 'Shroud' (by now internationally famous) was quite badly damaged by the melting of the silver box in which it was kept and the dowsing it received to stop it smouldering: the triangular repair patches done at this time are a prominent feature of the 'Shroud' still. It was rumoured at the time of the fire that the cloth had been destroyed, but let us accept that the cloth that went to Turin in 1578, with the patch-marks we see to this day, was the one that had come through Chambéry from Lirey. The cloth has remained in Turin (apart from a period during World War II) ever since 1578, when it was taken there from Chambéry to spare the archbishop of Milan (Cardinal, later Saint Carlo Borromeo) a journey across the Alps to make good his vow to visit the relic. The House of Savoy acquired the 'Shroud', probably in 1453 and in exchange for a castle and its revenues, not from the Lirey church direct but from one Marguerite de Charny whose father's home was at Lirey and whose family had actually owned the cloth since Marguerite's grandfather had, according to her, come by it as a spoil of war. This Geoffrey de Charny was a successful man of modest origins who followed a military career and was royal standard-bearer at the battle of Poitiers, where he was killed in 1356. Geoffrey had been a prisoner of war in England in 1350–1 and during this time he vowed to build a church if he was spared: hence the church at Lirey, begun in 1353 and consecrated in the year of his death. One year later, in 1357, an inventory of the Lirey church makes no mention of the 'Shroud', but it was being exhibited there very soon afterwards. Before the exhibitions, before the building of the Lirey church and before, if Marguerite was right about the spoils of war, the cloth came into the hands of Geoffrey de Charny, we cannot trace the 'Turin Shroud'.

Many writers have attempted to identify the Old Cloth of Lirey (which would be the best and least prejudiced name for the relic) with religious relics described by earlier mediaeval sources: in particular, with an object known as the Mandylion, which went missing from Constantinople when the Crusaders sacked the city in 1204 (at least 150 years before the appearance of the Lirey Cloth). Like the Old Cloth of Lirey, the Mandylion was by all accounts a piece of cloth bearing a human image – or part thereof, for the Mandylion carried only a bearded face and not a full-length figure front and back. Mediaeval paintings of it show it as a square with the head in the centre. According to legend, the Mandylion was a towel, used by Christ after washing, on which the image of his wet face became somehow imprinted. The Mandylion reached Constantinople in 944 from Edessa (modern Urfa in Turkey). It had been discovered there, so the story goes, during reconstruction work after a disastrous flood in the sixth century. A legend first recounted in the eighth century told that a king of Edessa, contemporary with Christ, had acquired the relic for his city. The identification of the Lirey Cloth with the disappeared Mandylion has been attractive to some 'Shroud' enthusiasts because, if correct, it would extend the history of the 'Shroud' back to the sixth century: but the gap between the appearance of the Lirey Cloth and the disappearance of the Mandylion is too great at a century and a half, the alleged nature of the Mandylion and its descriptions and depiction make it out to be nothing like the 'Turin Shroud' and, for full measure, there would still be more than 20 generations or so between the time of Jesus and the alleged discovery of the Mandylion at Edessa, with only a legend of later times to link the two. Incidentally, notions that the Mandylion was the 'Turin Shroud' folded up so that only its face was showing are quite unoccasioned and could only recommend themselves in the course of an effort to extend the story of the Old Cloth of Lirey back into the past at any price.

The Mandylion was not the only alleged image of Christ on cloth in circulation in the Middle Ages. Pilgrims to Rome believed that Christ's face was imprinted on the Veronica Handkerchief (which disappeared in 1527 when the army of Charles V looted the city). And the Emperor Charlemagne believed that he had acquired the true Shroud of Christ in 797 and this object, presented by his grandson Charles the Bold to the Abbey of Compiègne, survived until the French Revolution: but it had no image upon it. Another 'True Shroud', kept at Besançon, was shown in 1794 to be a painting. Yet another 'winding sheet' presented to Charlemagne by a contemporary caliph was found in 1945 to be decorated with lines from the Koran! At the very same time as they were displaying the Mandylion, the Byzantine authorities in Constantinople believed themselves to possess *the* 'True Shroud' too in the form of another relic, which a French Crusader saw, again in 1204. This witness, by the way, is the first to mention a shroud with an image on it: previously claimed shrouds had not been further alleged to carry images on them; only kerchiefs and towels did that. The 'Turin Shroud', then, is not unique though it has outlived most of its rivals, and more than one mediaeval priest believed that he was the custodian of the miraculous burial-image of Christ. Of course they cannot all have been right – were any of them? Nowadays, only the 'Turin Shroud' makes any widely-noted claim to be the Shroud of Christ, though as we have seen its history cannot be traced back beyond the fourteenth century.

The recent history of the Old Cloth of Lirey has been one in which it has moved out of the sort of benign obscurity in which holy relics have in general fallen into the light of enthusiastic investigation with modern more or less scientific techniques, with a view to establishing its authenticity as the Shroud of Christ. This process began in 1898 when, at the time of one of the cloth's periodic exhibitions, it was first photographed. The faint brown body-image upon a yellow ground was transmuted by the orthochromatic plate negative of the 'Shroud's' first photographer, using artificial light, into a much stronger and clearer delineation. Now in negative, the ground was dark and the body stood out from it in a light image that for the first time seemed full of detail and interest – the 'bloodstains', however, came out light too in this negative form. The photographic reproduction was widely circulated (the negative being much preferred to the positive) and provoked the first set of 'scientific' explanations of the image on the cloth. Among these was the 'vaporograph' theory of the turn of the century: in this notion, the image was produced by the action upon the cloth impregnated with myrrh and aloes of the vapour products of fermented feverish perspiration. This account of the formation of the cloth's image has since proved impossible to substantiate experimentally: the vapours from real bodies diffuse and produce no details at all.

In 1931 new photographs were made, which improved for clarity of image upon the first ones. These new photographs were also taken upon orthochromatic emulsion which is largely sensitive only to the blue component of light and not to the red end of the spectrum. Thereby the contrast between the light yellow cloth and the brownish body-image is increased and the body-image made to stand out more clearly. Recent photographs on panchromatic film disappointed their taker, with a much thinner and less distinct image: no doubt the first photographers of 1898 and 1931 were exposing, developing and printing for maximum contrast; their very clear productions have led some observers of the actual Lirey Cloth to be disappointed with the image before their eyes. The 1931 photographs created new interest in the cloth: a French doctor boldly experimented with cadavers to demonstrate the uselessness of nails through the palms as a method of crucifixion. With such an arrangement, the weight of the body was shown capable of tearing the hands off the cross. Nails through the wrists were more practical, as the Romans must have known, as a means of securing the body – and the 'blood' marks of the 'Shroud' image do indeed point to wrist wounds. Incidentally, the bones of a Roman-period crucifixion victim from Jerusalem, discovered in the 1960s, exhibit scratching of the radius bone in a way that seems to confirm the wrist theory of nailing. It has been claimed that in this respect the image on the Lirey Cloth displays an anatomical correctness about the details of crucifixion that was unknown to any potential forger during the Middle Ages. Constantine put an end to crucifying in the fourth century and artists have generally depicted Jesus with nails through the palms, but it is not true to say that wrist-nailing is unknown to European art: Rubens and van Dyck show it and it is to be seen on an ivory sent by the Knights of St John of Jerusalem to Pius XI.

Interest in the 'Turin Shroud', sustained by the 1931 photographs, continued in moderation with a small number of unshakeable enthusiasts until the early 1970s when, in 1973, the cloth was shown on television and a handful of scientists and experts were allowed limited access to it to make certain tests and publish results that greatly enhanced public interest. Speculations as to the mechanism of the creation of the image on the cloth by now included the plausible theory of scorching, perhaps in contact with a hot metal relief; the colour of the image and the known scorch marks of the fire of 1532 are very similar in appearance. The tests carried out in 1973 could not determine by what method the image had been created, but the report noted that the image was (apart from the 'blood' marks) confined to the surface of the cloth. One of the experts believed that the image might have been *printed* onto the cloth from another one, and even speculated about a fifteenth-century date, relating the technique to some of the painting methods of Alberti and Leonardo. More importantly, an analysis of the pollen traces on the Lirey Cloth was judged to indicate a history that included not just north France and Italy but also Turkey and Palestine: it was concluded that 13 of the 48 plants that could be identified by pollen on the cloth could be tied to the area around Jerusalem at the time of Christ; these pollen analyses have not been repeated by other researchers and have been described by one palynologist as 'incredible'. The 1973 team concluded that the actual weave of the cloth was possibly consistent with, but not positively indicative of, a context of first-century Palestine: the same sort of cloth has been made over a long period since at least the second century AD. As to whether the Lirey Cloth conforms to the Gospel accounts of Jesus' mode of burial, where he is described as *wrapped* in the burial-cloth (the 'Sindon' of Matthew, Mark and Luke) or cloths ('Othonia' of John) it appears that the Gospels are sufficiently imprecise as just to permit the correlation. But if the front-and-back figures on the Lirey Cloth were formed in proximity to an actual body, it has to be said that the body cannot have been in very close contact with the cloth all over or a highly distorted image would have been produced (by whatever means): all the more or less natural explanations like vaporography from sweat fall foul of this objection. The poncho-like configuration of the Lirey Cloth necessary to produce its image (again by whatever means) in the course of use as a burial-sheet is not attested in Jewish funerary tradition.

In 1977, the study of the Lirey Cloth took on what some commentators are pleased to call a 'space-age' character with the researches of two American airforce 'Shroud' enthusiasts. They reasoned, taking up the observations of others before them, that the undistorted image on the cloth does not seem to have been produced *in full contact* with a three-dimensional object. They claim that the density of the image on the cloth can be correlated with a cloth-to-body distance, with the densest parts nearest (and touching) the body and the faintest parts furthest from it. Making the assumption that the image on the cloth was somehow formed by its locally varying proximity to a human body, they selected a male model of the right sort of size and draped a replica of the cloth over him in a manner they imagined it was plausible to think a cloth would have lain over the body of the original subject: measurements of cloth-to-body distance were then correlated with image density. With recourse to a 'V.P.-8 Image Analyser', the airforce captains created on the basis of their figures a sort of three-dimensional version of the 'body' whose image is on the Old Cloth of Lirey which has impressed a lot of journalists, notwithstanding that the face and the body display such different characteristics that, when the face is about right for depth, the body becomes something like a bas-relief and that the 'wounds' on the 'Shroud' image show no depth to them when thus rendered in 3-D. Their work, moreover, is based upon photographs of the 'Turin Shroud' which modify the contrast-range of the original article. Rather in the manner of the psychologists' ink-blot test, some enthusiasts have even been able to spot, in a lump on the forehead of the three-dimensional picture, some sort of phylactery (Jewish prayer-box) and even a coin over one of the eyes of the time of Pontius Pilate! It may well be that, like the 'image-enhanced' photograph of the Loch Ness Monster's flipper published in the papers a few years ago, a theory and its associated technique have been wholly misapplied to an object whose nature has nothing to do with the assumptions of that theory and technique.

The airforce men's work inspired a new spate of interest in the Lirey Cloth, with a pronouncedly American and 'high-technology' slant. Each and every scientific test that might be applied to the cloth was canvassed by the American group and the results were interpreted in a spirit ranging from the usefully open-minded to the senselessly speculative. For an example of the latter, we can do no better than quote the words of one enthusiast: 'I believe both images were caused simultaneously when laser light revived the lifeless body in the shroud, triggering off a thermonuclear explosion that lasted for a millisecond of time. The body was transformed molecularly which enabled it to pass through the shroud, while radiant heat scorched the negative image onto the cloth in a precise and controlled manner . . .'

In 1978 the Cathedral authorities accepted all the proposals for scientific tests put forward by the American-inspired investigators, bar one – the test for radiocarbon dating, which could settle the matter of the 'Shroud's' plausibility as the burial-cloth of Jesus once and for all, was not allowed on the grounds that too much of the cloth would have to be sacrificed to achieve the date. This was a disappointment and efforts have since been made to convince the authorities that new methods of radiocarbon dating, involving much smaller samples than previously, could now be applied. The results of the other tests that were carried out following upon the authorities' permission of 1978 have revealed a division of opinion as to their interpretation. One researcher has concluded that there is clear proof of the creation of the image on the Lirey Cloth by means of an iron-oxide pigment on a sized base, with the addition of some other pigments typical of an artist's studio – in particular in the 'blood' stains, where vermilion has been detected. Dr McCrone, who has to his credit the debunking of the Vinland Map (a forgery which purported to show the East Coast of America at a time before Columbus crossed the Atlantic), has said of the 'Shroud' that 'the partially negative image with its apparent relationship to a three-dimensional figure is a natural result of an attempt to paint the image as it might be expected to register on a cloth covering a dead body'. Others among the Americans who have carried out tests on the cloth disagree with this verdict and account for the undoubted presence of iron-oxide as a result of the processes of cloth-manufacture in the past, or as the result of the breakdown of components of real blood which some of them think is present in the 'blood' stains. Dr McCrone replies that a person would have to bleed iron-oxide to leave so much of it in the image of the cloth. There may be real blood present in the 'blood' stains (whether human or animal) but there is no doubt that the 'Shroud's' 'blood' stains are both too red and too picture-like in their disposition to be blood and only blood, and they must at the very least have been touched up with artist's materials. They have not, moreover, dried like real blood on a cloth that has been in contact with a bleeding body. The use of some real blood in the 'blood' stains on the Old Cloth of Lirey would not, of course, be beyond a forger.

Dr McCrone's opponents in America have speculated wildly about the creation of the image on the 'Shroud', frequently returning to the brief-flash-of-intense-radiation theory that cannot help but hint at supernatural intervention. One ingenious notion, relying upon the ageing of cloth imprinted with sweat and spices, even envisages that the 'Shroud' (although carrying as it were a latent image) did not come to manifest its figure until the Middle Ages; and there is a separate strain of speculation, quite beyond notions of how the image was formed, that wonders whether it represents a dead or only catatonic subject – we have been categorically assured both that the 'blood' stains could only have come from a still living and circulating body and that the figure shows all the signs of rigor mortis! But what all the enemies of any explanation of the 'Shroud' image in terms of human artifice endlessly assert, is that no forger could have painted the figure, however 'cunningly' to mention bishop Pierre's allegation again. Indeed they point out, quite correctly, that the image shows none of the typical traits of any sort of painting, with – for instance – no directionality of brush marks. But it must be said that other methods of image-formation with pigments are entirely possible, such as printing from one cloth to another or from a paint-loaded statue or relief.

In fact, two experiments have suggested that it might be possible to produce the image on the Lirey Cloth by means available to an artist-forger of the Middle Ages. A modern artist recently painted a face reminiscent of the one on the 'Shroud' on a piece of sized cloth, puting more paint on at points where she thought her cloth would have contacted an imaginary body beneath it and less and less with distance away from the cloth. A forger of the Middle Ages, trying to paint a shroud-image that depended on contact with the deceased in the darkness of the tomb rather than on the play of light on a viewed body, might have been just as clever (indeed, in a commercially superstitious age he might have had reason to be cleverer). The modern artist's painting not only strongly resembles the 'Shroud' face but, when subjected to the same cloth-to-body image-density correlation as the airforce men applied to their photographs of the 'Shroud', produced a satisfyingly three-dimensional image too, much like the 'Shroud's'. In another experiment, a piece of cloth was wet-moulded over a bas-relief carving of a face not unlike the 'Shroud's' and, when dried in place on the relief, was impregnated with a spice-concoction by means of a porous dauber. This experiment, too, produced an image very like the one on the 'Shroud', with no signs of directional brush-work, of course, and with crease-marks (caused by the moulding of the cloth) closely resembling the lateral marks seen clearly across, for instance, the chin of the face on the 'Shroud'. It seems likely that a further experiment in which a piece of cloth was sized and moulded to a bas-relief carving (bas-relief rather than fully in-the-round, because the latter would introduce distortions), then daubed with a semi-solid iron-oxide pigment and finally washed (as we think the 'Shroud' was piously washed in the Middle Ages) might produce an image not just visually resembling the faint image of the 'Shroud' but chemically resembling the composition of that image too. 'Cunningly painted' indeed! So it is not true to say, at this stage, that the image on the 'Shroud' could not be a fake made by human hands. Such fakery would at least explain why the right forearm of the figure on the 'Shroud' is too long and the fingers of the right hand too extended – artists do that sort of thing!

In the absence of radiocarbon dates, we can only conclude that there is nothing about the image upon the Old Cloth of Lirey that is inconsistent with a French Gothic origin in the fourteenth century AD. The cloth cannot be historically pursued beyond this period and the image itself is quite consistent with the French Gothic in general angularity of features and in details of hair and beard: indeed, in the Tuchlein paintings on cloth of the end of the thirteenth century, there are parallels for it in technique too. And this was a period of intense interest in religious relics (both sentimental and commercial) when the more morbid aspects of the image of Christ Crucified came to the fore in Christian art and bands of Flagellants and other religious fanatics were infesting Europe with their obsessional mortifications modelled on Christ's sufferings on the Cross. Of course, the significance of the 'Turin Shroud's' relationship with French Gothic (or any other) art and the context of fourteenth-century Europe (or any other period) would assume greater interest for the art-historians if a positive date could be assigned to the cloth. For them, one might say, it would be of great interest if demonstrated to be of fourteenth-century date, or even greater interest if shown to be of – say – seventh-century date, and of greatest interest of all for art-history and the history of Christian iconography if proved to belong to the first century AD. It should be noted, incidentally, that available examples of the early iconography of Christ, dating to the third century AD, do not represent Christ Crucified (images of crucifixion appear in the next century) but render Jesus in the guise of the Good Shepherd, in an idealised manner strikingly similar to the contemporary pagan depictions of figures like Apollo and Orpheus. At the same time, the tendency of Christian writers of the late second and early third centuries AD to describe Jesus in physically unflattering terms as ill-favoured and even ugly in appearance, seems to have derived from their wish to identify him with Isaiah's 'suffering servant' of the Old Testament. The Gospels, written at intervals during the latter part of the first century, do not describe the appearance of Jesus and there are no independent texts by any Jewish or Roman writers that prove him to have existed at all.

For historians and archaeologists, the Lirey Cloth would be comparatively interesting as a fourteenth-century fake, more interesting as a seventh-century fake and very interesting indeed if it appeared to belong to the first century, fake or no. Whilst it is never likely to be capable of such proof, the Old Cloth of Lirey would be overwhelmingly interesting to the historians (among others) if it could be proved to be the authentic Shroud of Christ – for it would be the only independent corroborative evidence of the historical existence of Jesus of Nazareth (of course, a date of 200 or 300 years BC would be even more interesting in its way).

It is not likely that such proof of the 'Shroud's' authenticity as the burial-cloth of Jesus can ever be obtained, and those who pin their faith on radiocarbon dating as a means by which that authenticity might be demonstrated, are sometimes deluding themselves as to its possibilities with regard to the Lirey Cloth. If a radiocarbon date of AD 1300 becomes available, then we shall know that the 'Shroud' is a fraud, of a sort much as we might reasonably have expected. If a date of AD 700 comes along, then we shall know equally that the cloth is a fraud – this time of a more surprising sort and relating to perhaps unexpected milieus of perpetration. But what of a radiocarbon date for the 'Shroud' of the first century AD? Say something like, to make the case vivid, AD 30 $\pm$ 50 years?

Here it is necessary to recall something about radiocarbon dates often forgotten even by archaeologists: the manner of recording a radiocarbon dating determination contains a 'margin of error' as it were – in this case, those $\pm$ 50 years. The figures quoted in plus-and-minus terms are based upon the 'standard deviation' whereby there is about a 68 per cent likelihood that the real date of the sample falls somewhere between the earliest end of the range (in this case 20 BC) and the latest (AD 80) *with no preferences for any particular year within that range.* The apparently central figure of AD 30 in our example would share the same 68 per cent likelihood of being the true date with all the other years between 20 BC and AD 80. To increase the likelihood to 95 per cent, we would need to double the standard deviation and talk about a range from 70 BC to AD 130. In other words, the sort of radiocarbon date we are likely to obtain for the Lirey Cloth, even in the event of its coming out at somewhere in the first half of the first century AD, would leave us with an item that might range at least from the early years of Augustus to the rule of Domitian, with no preference for the time of Tiberius and Pontius Pilate. Even having said that, we should only know that the plants out of which the Lirey Cloth is woven died somewhere between at least 20 BC and AD 80; we would still not know when the image got onto the cloth, it being patently obvious but sometimes overlooked that the cloth might be old and the image still belong to the fourteenth century AD, since we shall be dating the cloth and not the image it carries. It is for reasons of this sort that archaeologists like to have available a range of material to date that is associated with the period or culture or individual topic they are studying, and thus obtain a range of overlapping radiocarbon dates both for corroboration and refinement of dating.

So radiocarbon dating can only solve the problems of the Lirey Cloth that exercise its would-be believers and detractors by conclusively proving – with a late date – that the 'Shroud' is a forgery and thus disposing of it for those without an art-historical interest in it or an interest in the by-ways of forgery and credulity.

Further Reading

The Evolution of Man

John Reader, *Missing Links*, Collins 1981

Desmond Collins, *The Human Revolution*, Phaidon 1976

M. Gerasimov, *The Face Finder*, Hutchinson 1971

Origins of Civilisation

Jacquetta Hawkes, *The Atlas of Early Man*, Macmillan 1976

Cambridge Encyclopedia of Archaeology, Cambridge University Press 1980

Encyclopedia of Ancient Civilisation, Windward Books (W.H. Smith) 1980

Egypt and Mummies

Rosalie David (Ed.), *Mysteries of the Mummies*, Cassell 1978

Paul Jordan, *Egypt, the Black Land*, Phaidon 1976

James Harris and Kent Weeks, *X-raying the Pharoahs*, Macdonald 1973

Bog Bodies

P.V. Glob, *The Bog People*, Faber and Faber 1977

Geoffrey Bibby, *The Testimony of the Spade*, Collins-Fontana 1962

D.W. Harding, *Prehistoric Europe*, Elsevier-Phaidon 1978

The Classical World

Peter Levi, *Atlas of the Greek World*, Phaidon 1980

Barry Cunliffe, *Rome and Her Empire*, The Bodley Head 1978

Raleigh Trevelyan, *The Shadow of Vesuvius*, Michael Joseph 1976

The Turin Shroud

Ian Wilson, *The Turin Shroud*, Gollancz 1978

H. David Sox, *File on the Shroud*, Coronet Books 1978

Index